Safety First!

This book belongs to ____________________

Name ______________________________ Cut and paste.

Some things are not for eating.

Safe

Not Safe

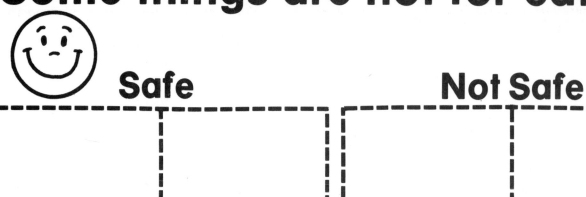

Name ______________________ Cut and paste in order. Color.

Wear your seat belt.

Name ____________________ Color. Write the phone numbers.
Keep this by your phone.

Emergency Numbers

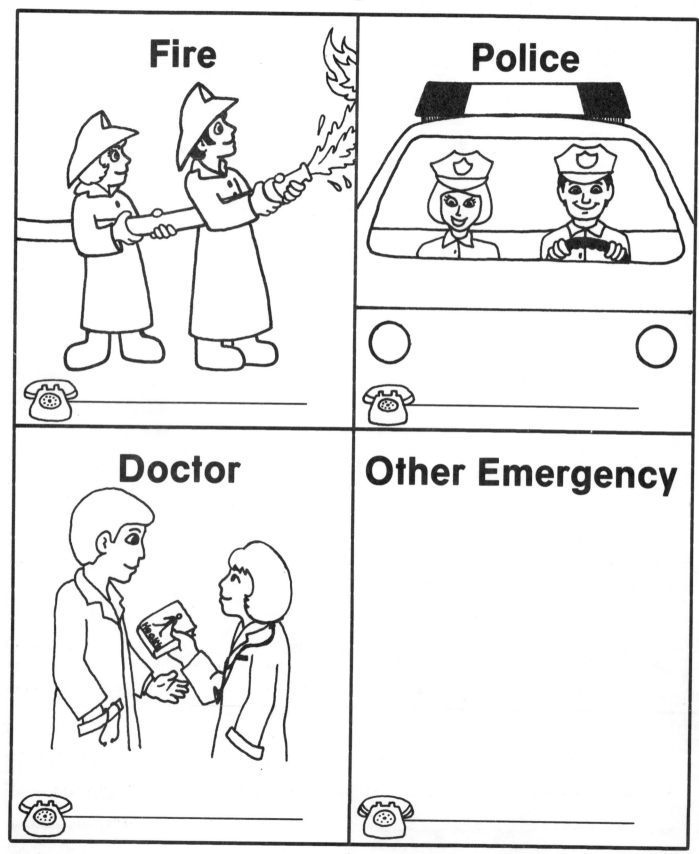

Fire

 ____________________

Police

 ____________________

Doctor

 ____________________

Other Emergency

 ____________________

Name ______________________________ Color, cut and paste.

Play It Safe

Never go with strangers.	Stand still. Say, "Go home!"
Put things where they belong.	Never play with matches.

Name ________________________________ Draw lines to match the places and signs.

Signs to Know

DANGER

KEEP OUT

EXIT

WATCH YOUR STEP

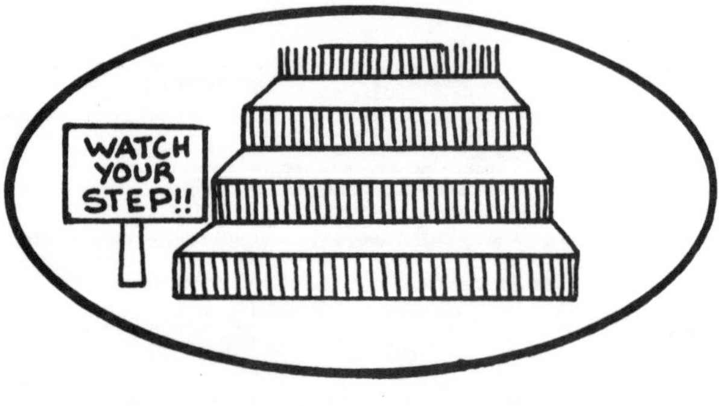

KEEP OFF

Name ______________________ Find the safety message.

B ___   ___ ___ ___ ___   ___ ___ ___

5 8   6 9 3 8   4 2 8

___ ___ ___ ___   ___ ___   ___ ___ ___!

10 7 3 8   7 6   1 9 4

Name ______________________________ Cut and paste.

Use Electricity Safely

Safe | **Not Safe**

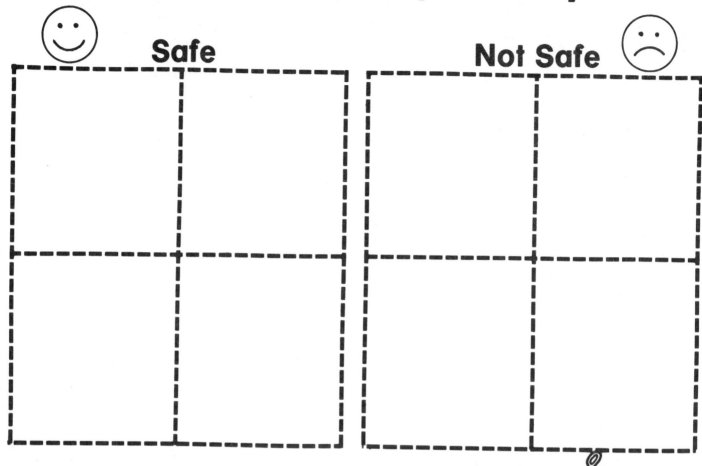

5 SENSES

This book belongs to: ______________________________

Name ______________________________ Color the things you might like to taste.

TASTE

Name ______________________ Color. Cut. Paste in the correct column.

SEE **HEAR**

SEE	HEAR

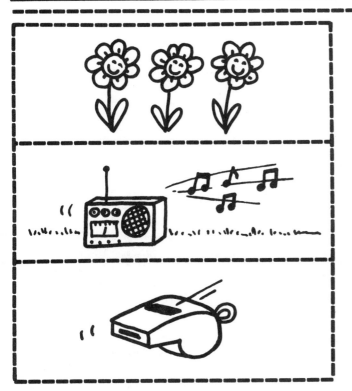

Name ______________________________

Color. Cut. Paste in the correct column.

SMELL

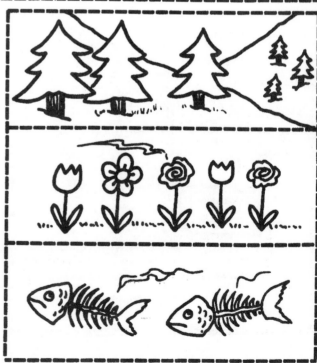

Name ______________________

Help Spot track the bone.
Follow the maze, color.

SMELL

Some animals have a good sense of smell!

Name ______________________

Color. Cut and paste squares in the correct columns.

HOT

COLD

TOUCH

Name ______________________________ Match the 5 senses.

My Book About Healthy Teeth

This book belongs to ______________________________

Name ______________________________ Color a square each day you brush your teeth.

I Brush My Teeth Regularly

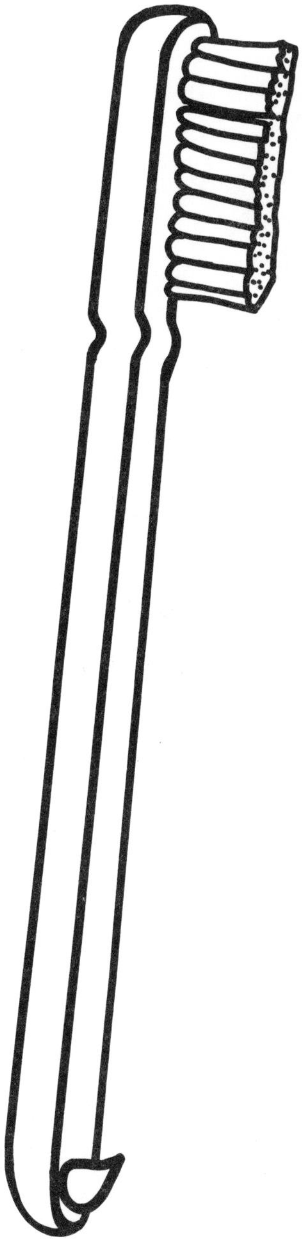

S	M	T	W	Th	F	S

Name ____________________ Cut and paste to match the pictures.

I Brush My Teeth Every Morning

I eat breakfast.	**I put on my coat.**
I get dressed.	**I brush my teeth.**

Name ______________________ Match the word to the picture.

Dental Tools

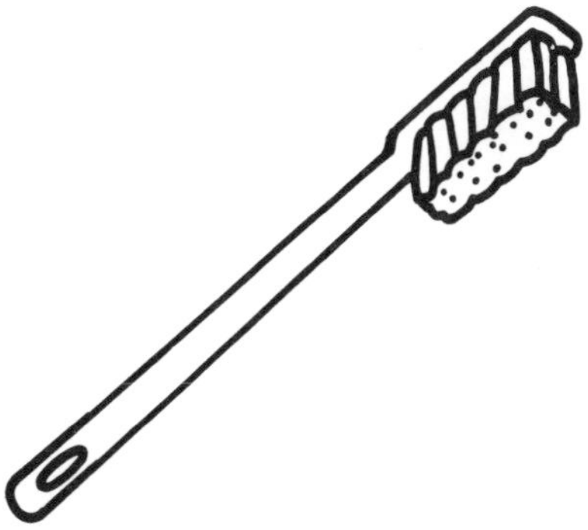

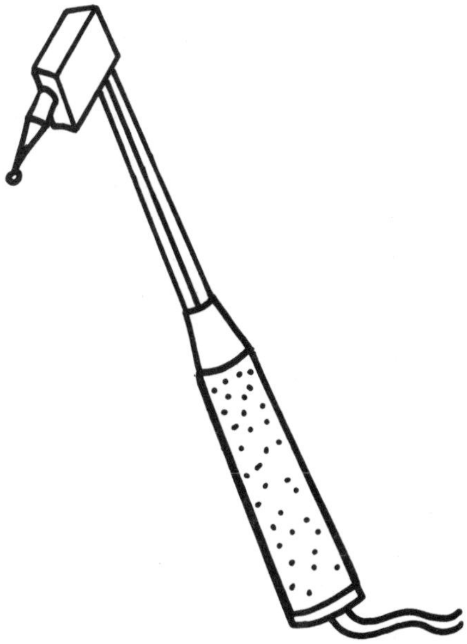

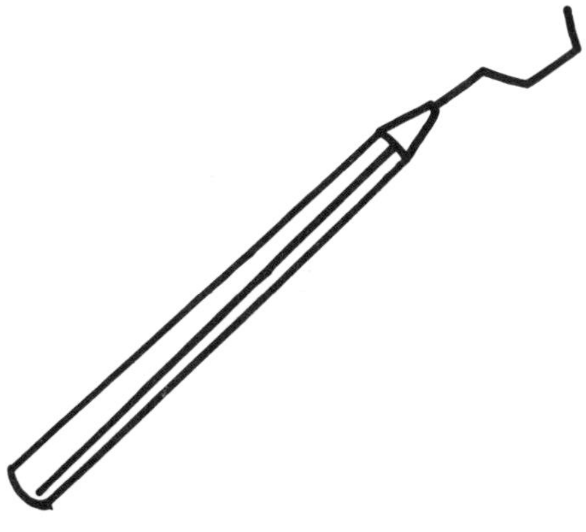

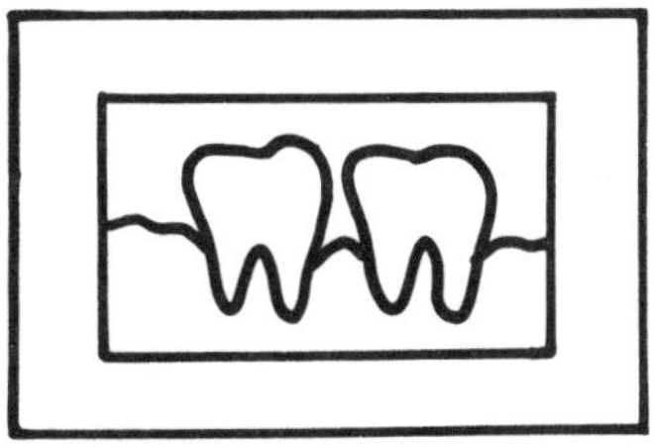

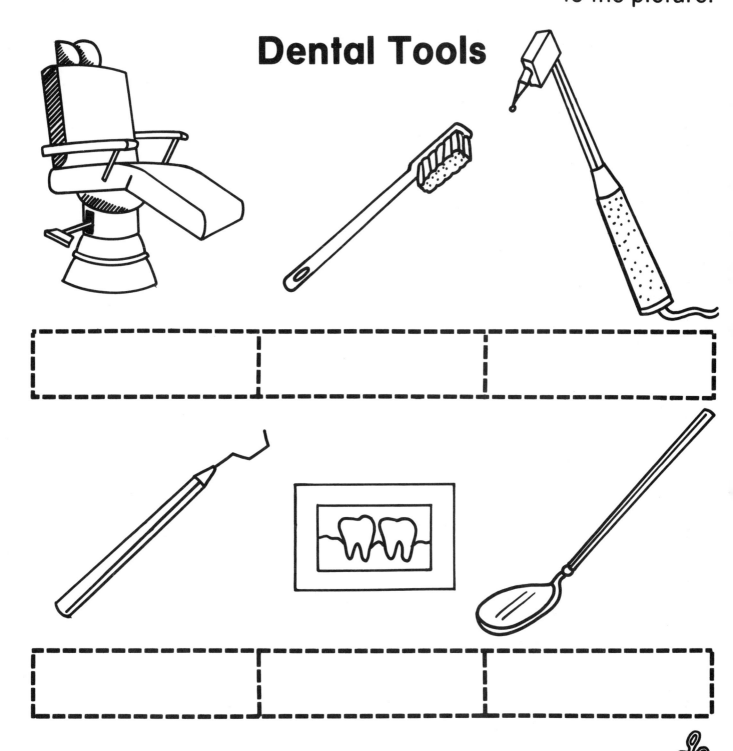

drill	**chair**	**explorer**
mirror	**toothbrush**	**x-ray**

Name ______________________________ Cut and paste in order.

I Visit My Dentist Regularly

1	2
3	4

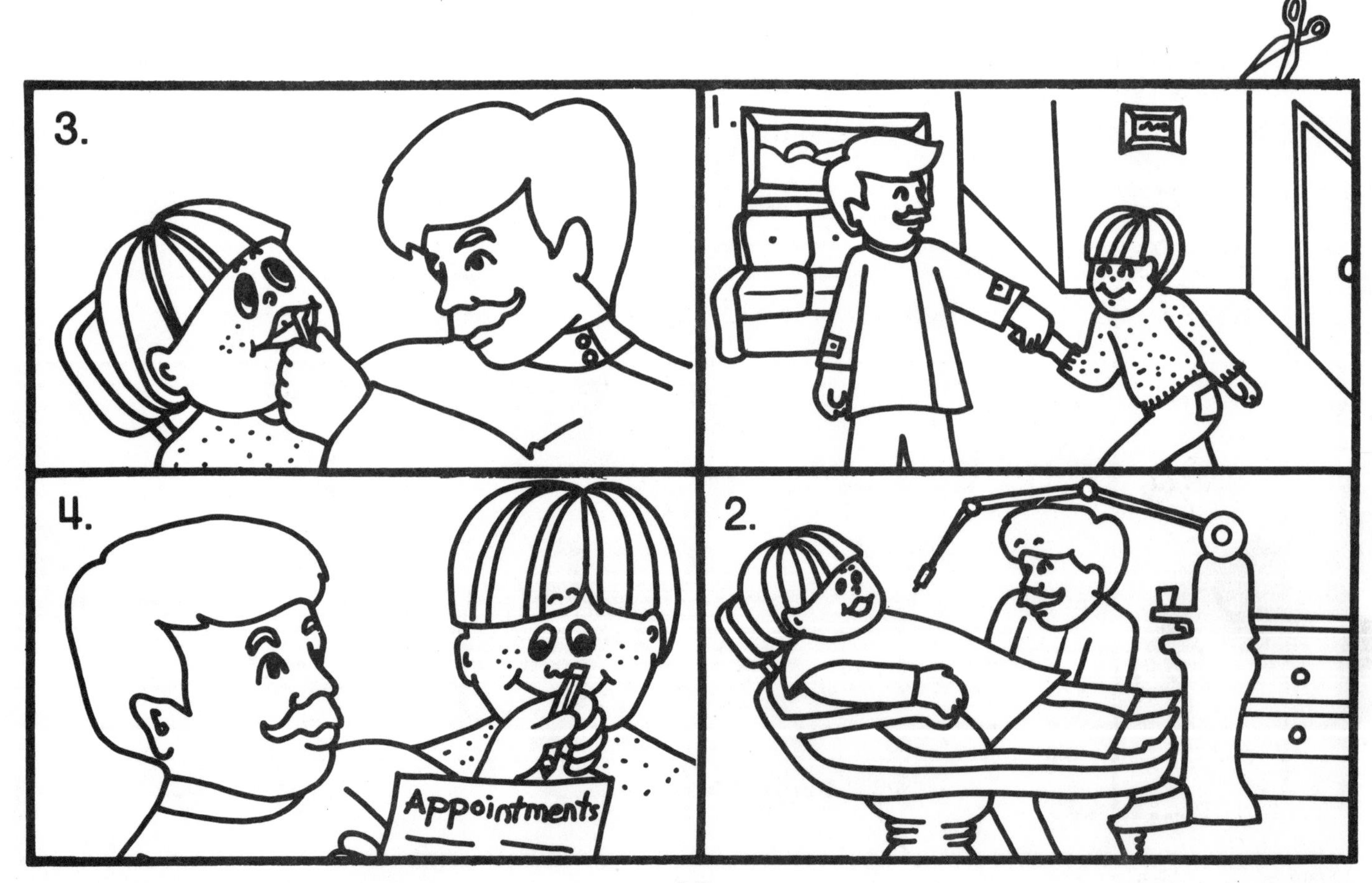

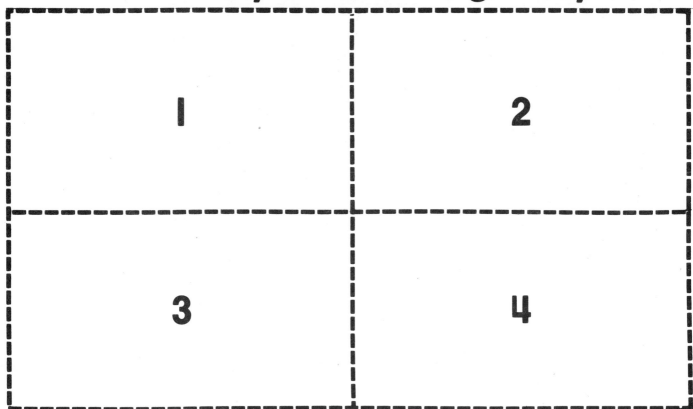

Name ______________________________

Color the healthy snacks to make a path.

I Choose Healthy Snacks

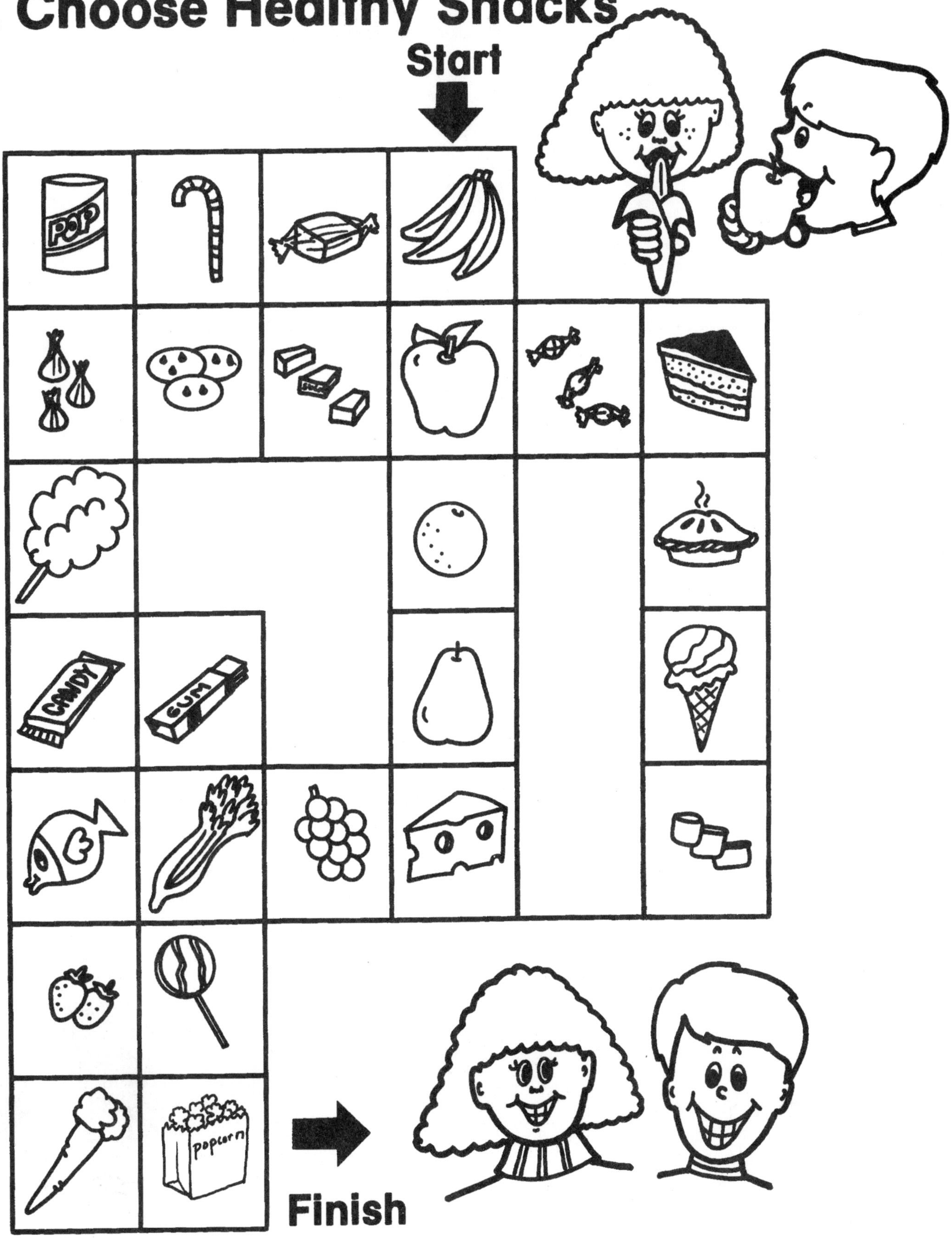

Name ______________________________ Count the number of primary teeth.

The First Teeth Are Primary Teeth

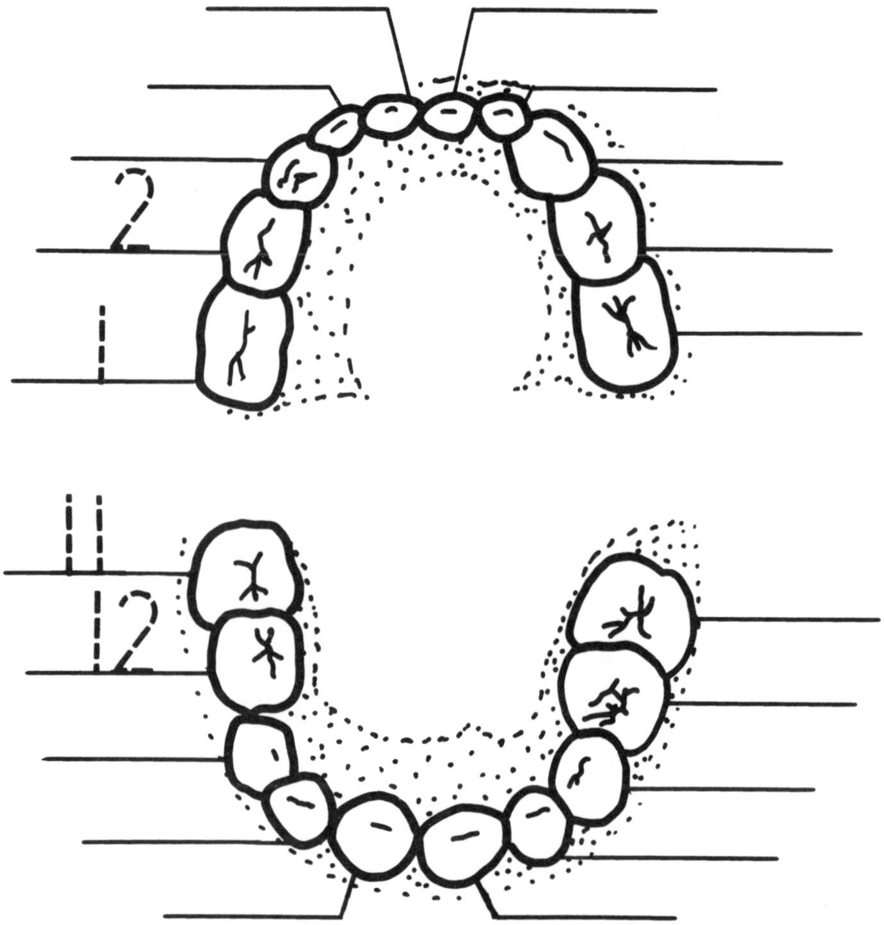
2
1
11
12

How many teeth do you have?

There are ________ primary teeth.

I have ________ teeth.

The Human Body

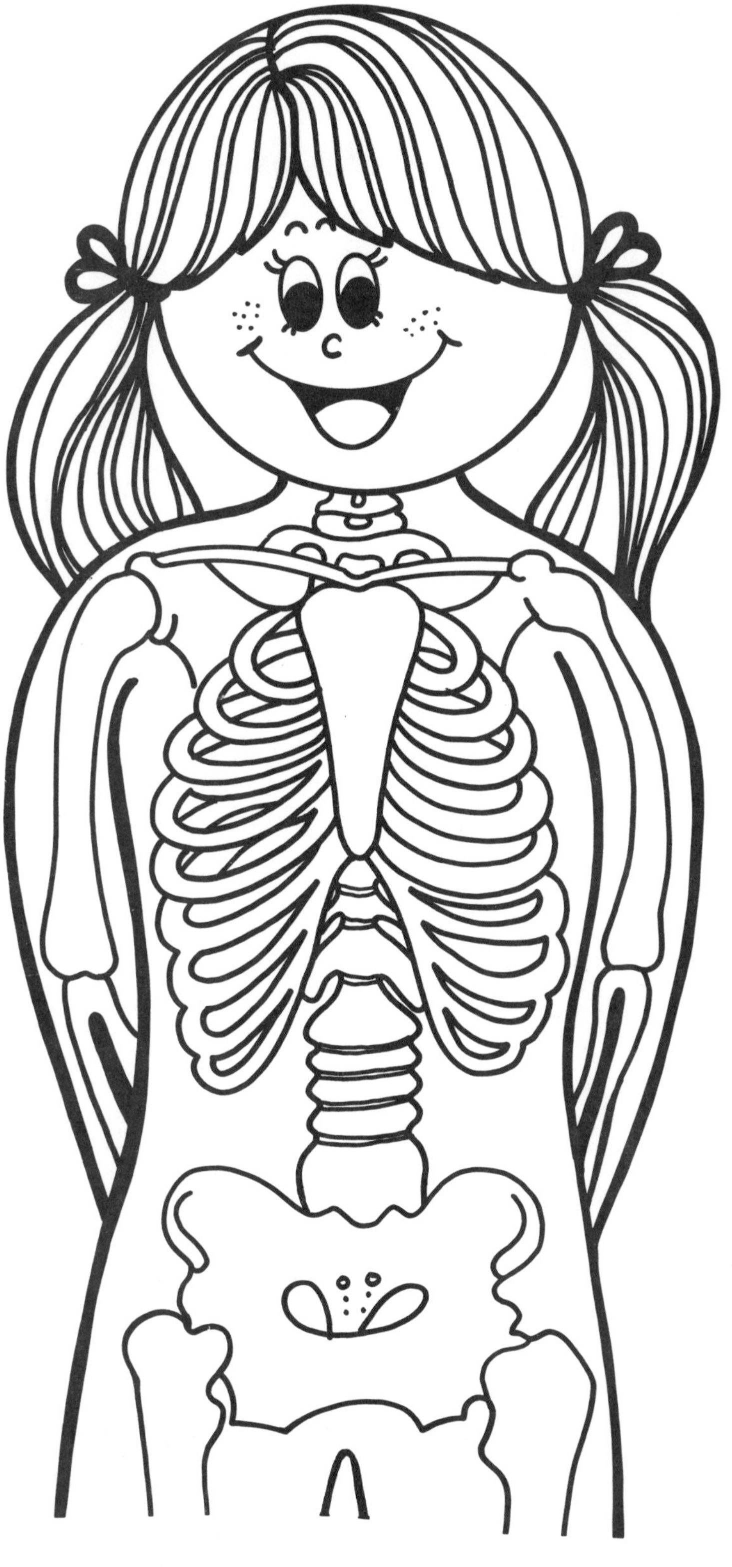

This book belongs to ______________________________

Name ______________________

Weigh, measure and draw yourself.

My Body Grows

When I Was Born	Now
I weighed ________	I weigh ________
I was this tall: ________	I am this tall: ________
My picture then:	My picture now:

Name ________________________ Cut and paste the pictures.

Food Is Digested

Name ______________________ Cut and paste in order.

My Body Has Muscles

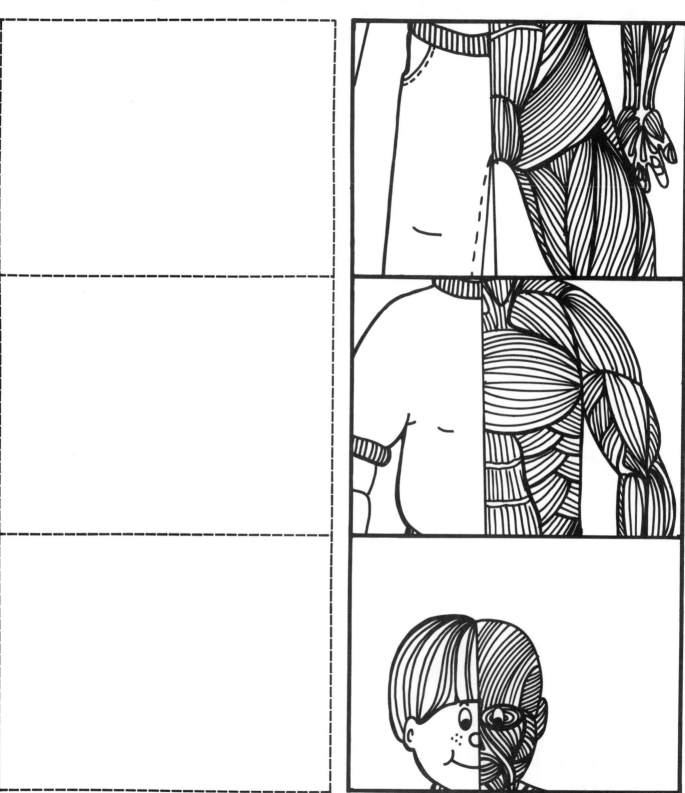

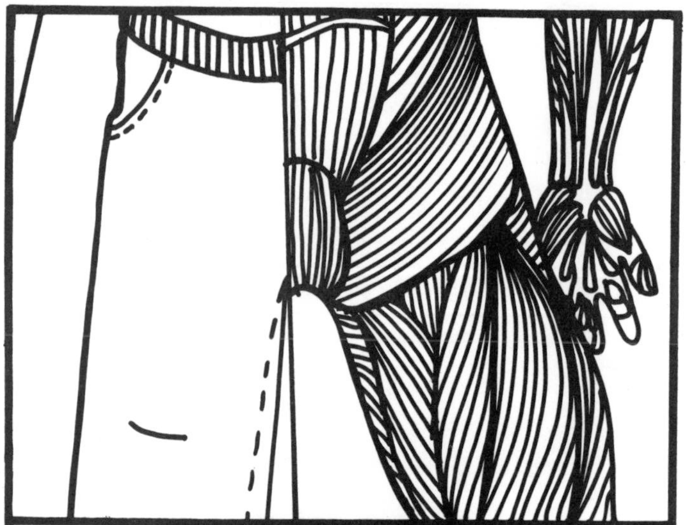

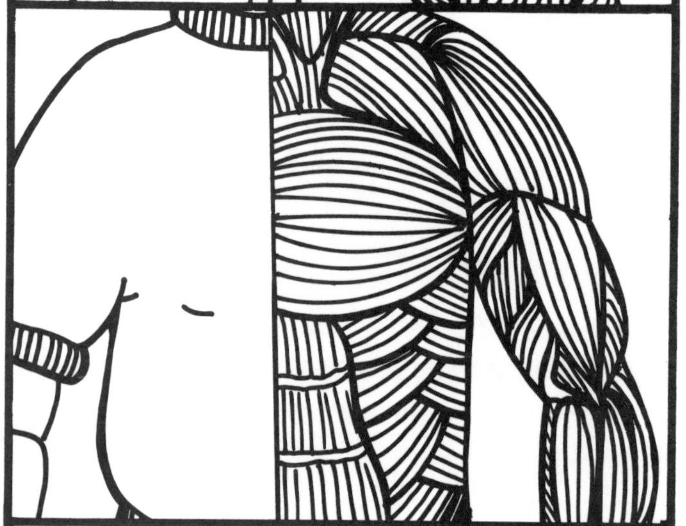

Muscles help me move.

Name ______________________________ Cut and paste to label the bones.

I Have Bones

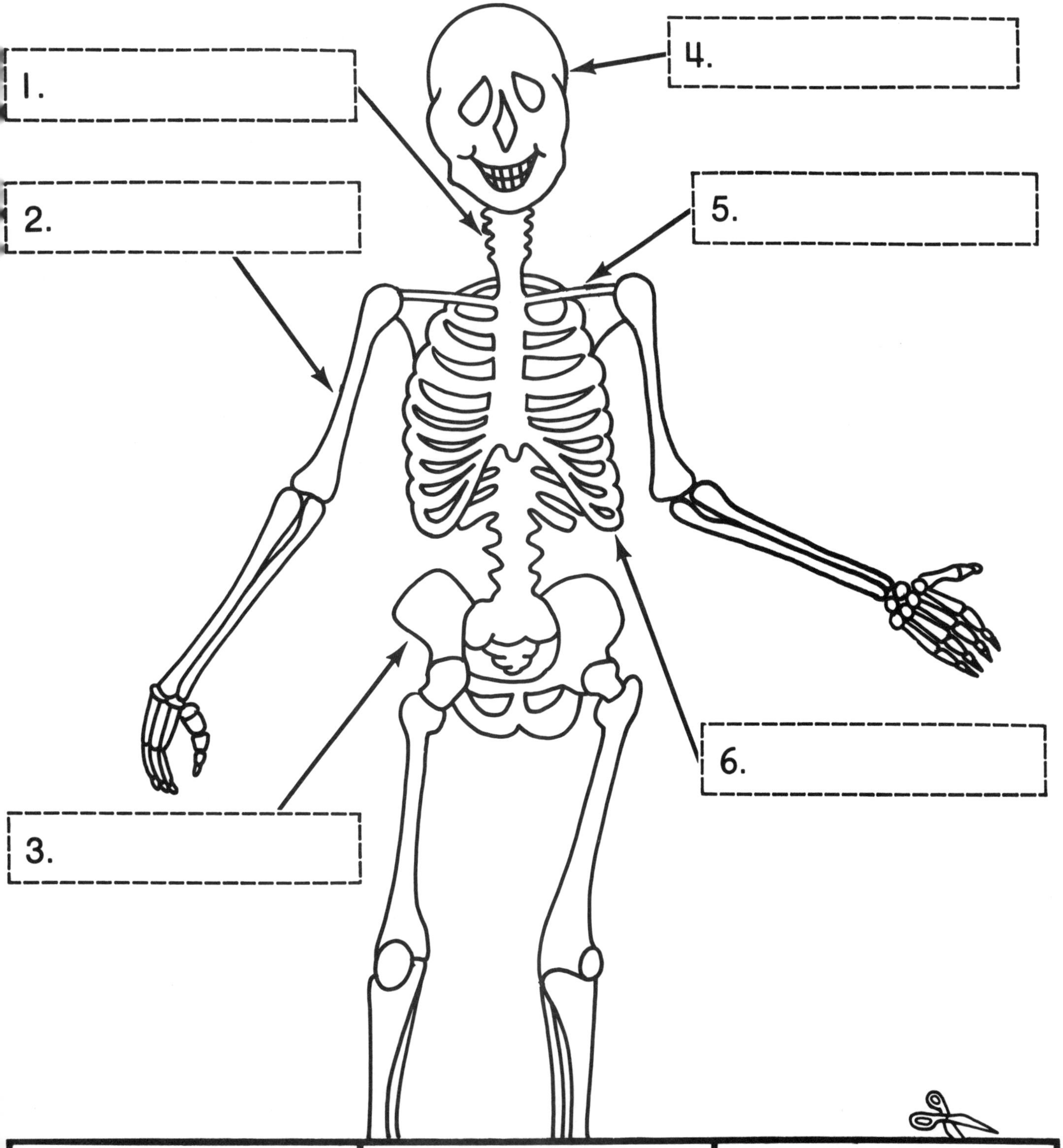

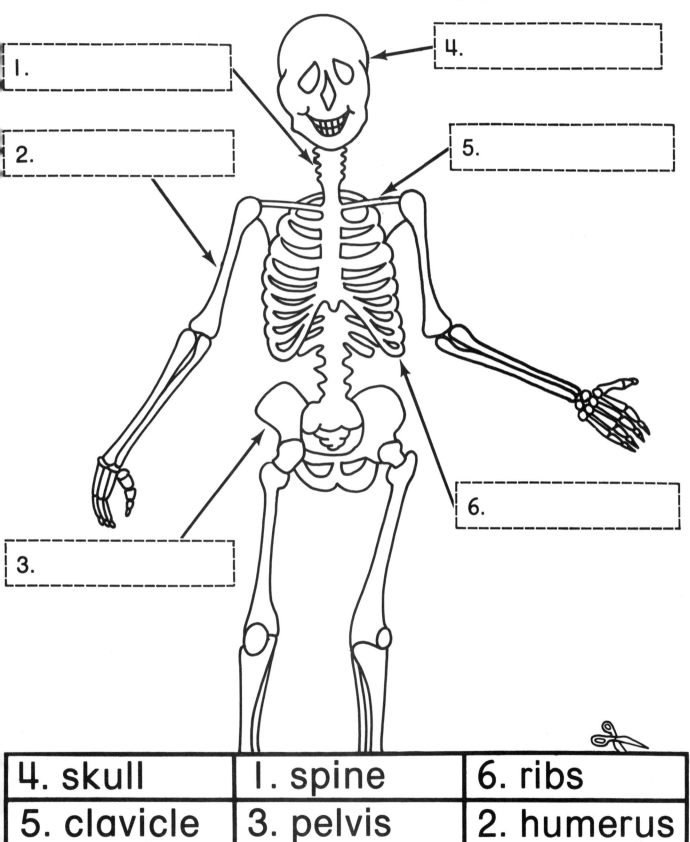

4. skull	1. spine	6. ribs
5. clavicle	3. pelvis	2. humerus

Name ____ Draw pictures in the correct box.

I Care for My Body

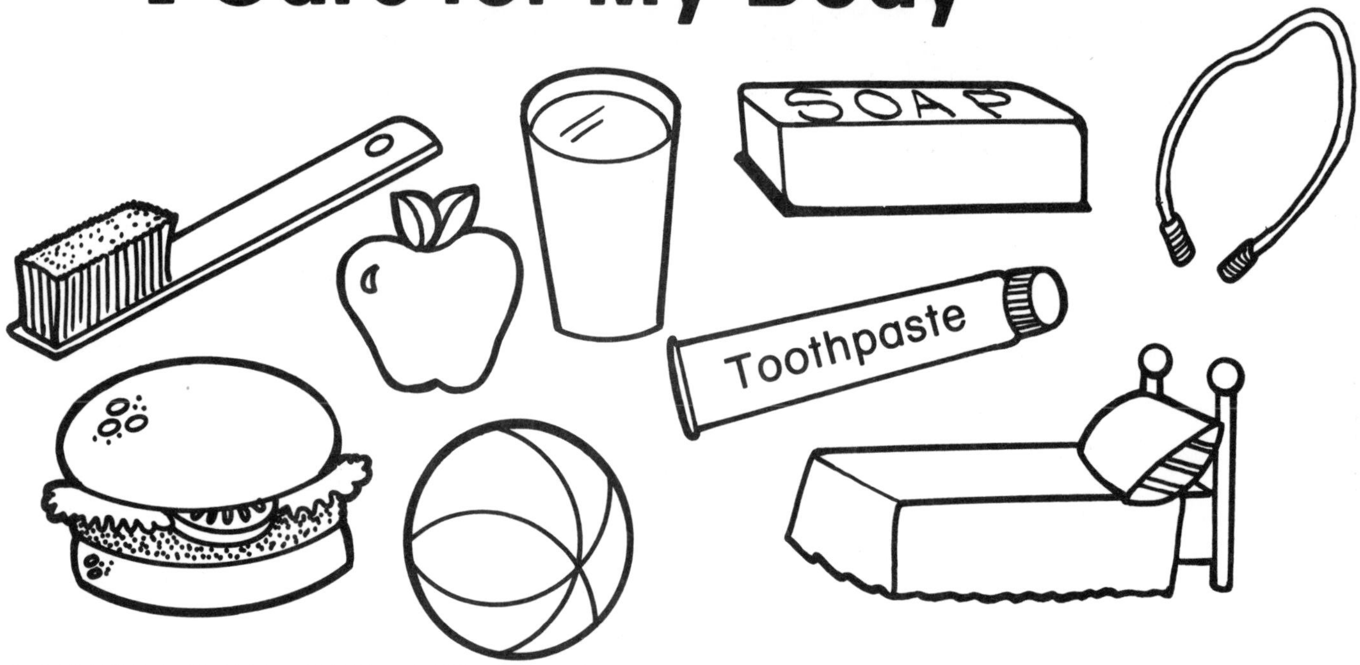

I exercise and rest.	I eat good food.	I keep myself clean.

Learning About Trees

Honey Locust

This book belongs to ______________________________

Name ______________________________ Color, cut and paste.

From Seed to Tree

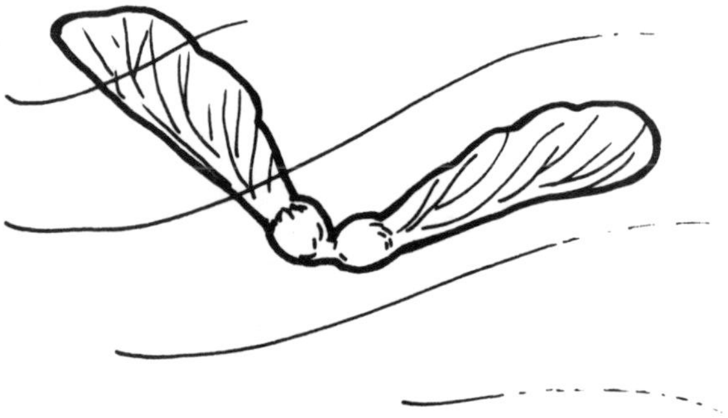

1.

2.

3.

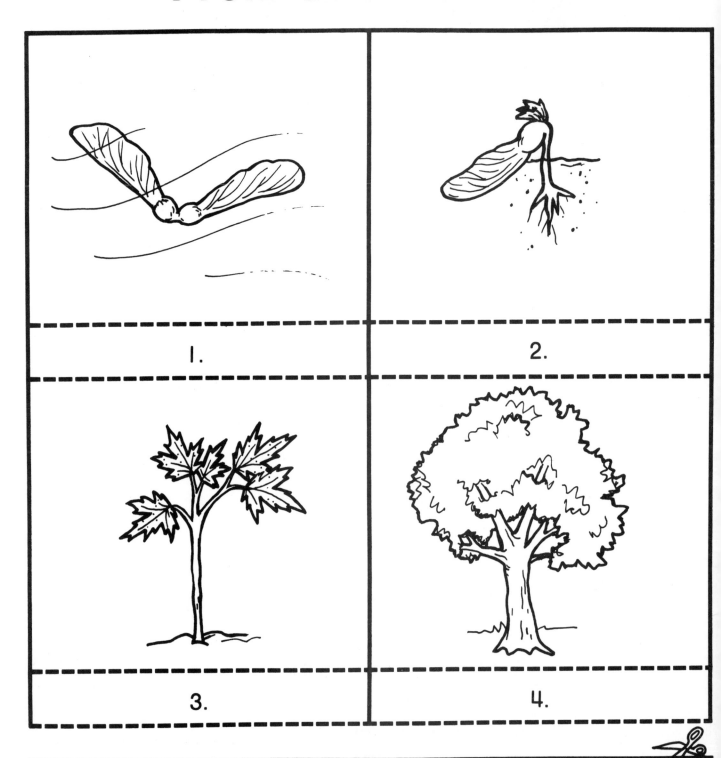

4.

2. The seed begins to grow.	4. It grows into a maple tree.
3. It becomes a sapling.	1. The wind carries a seed.

Name ______________________________ Color, cut and paste.

Parts of a Tree

trunk	branch	leaves
flowers	fruit	roots

Name ______________________________ Color and remember.

Leaves make food for trees.

Maple

Oak

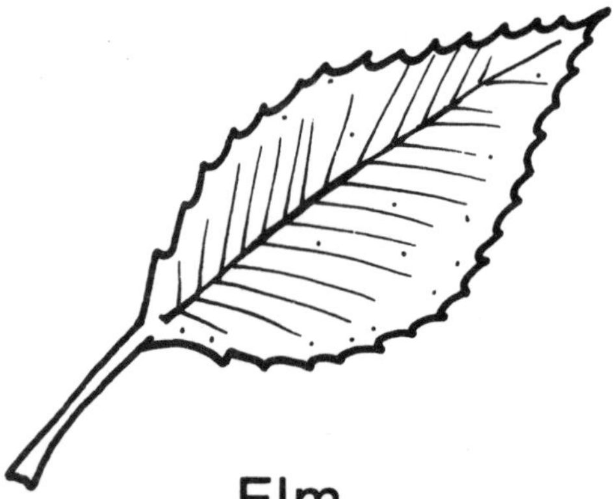

Elm

Pine

Cottonwood

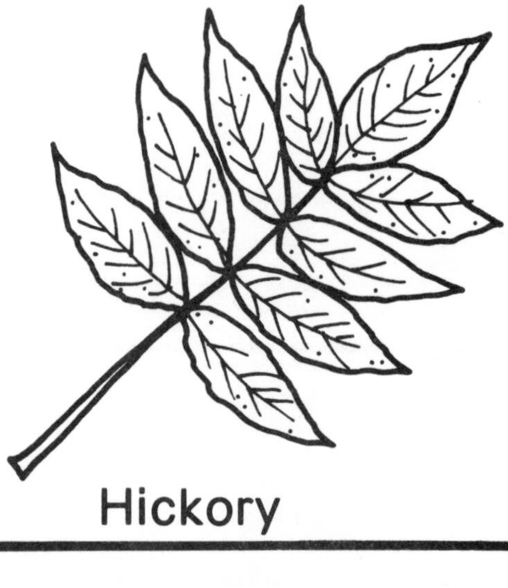

Hickory

Name ______________________________

Draw a line to connect the matching trees. Color.

Shapes of Trees

Name ______________________ Color and remember.

Some leaves change color.

Name ______________________________ Color the gifts trees give us.

Trees give us many gifts.

Flowers

This book belongs to ______________________________

Name ______________________ Color, cut and paste.

Parts of a Flower

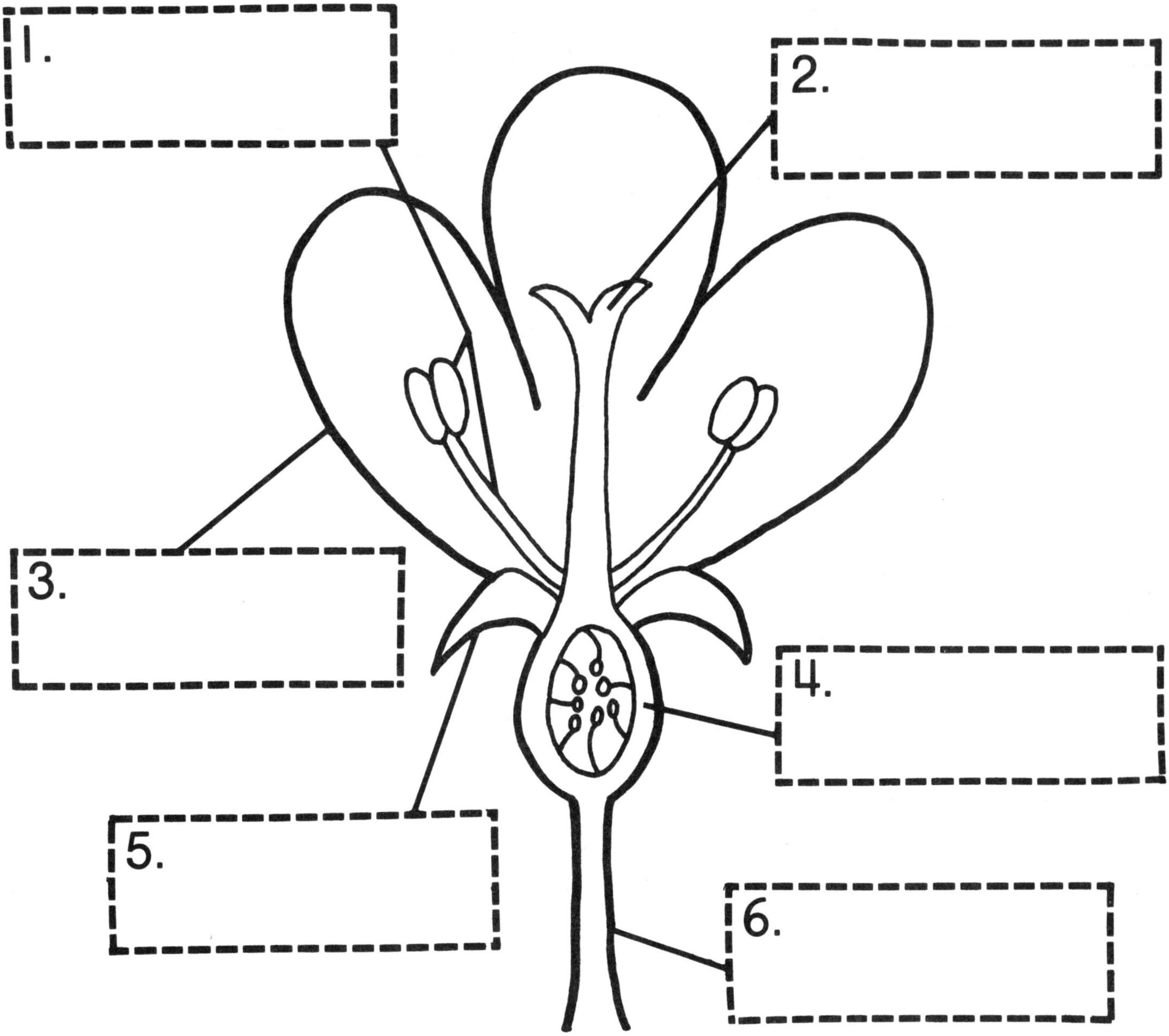

5. sepal	3. petal	2. pistil
1. stamen	6. stem	4. receptacle

Name ______________________________ Trace the lines. Color.

Insects Pollinate Flowers

bee

red clover

butterfly

sunflower

moth

honeysuckle

wasp

fig

Name ______________________ Color and cut. Staple to make a book.

Wild Flowers Are Colorful

Wild Flowers

Name ______________________

Lupine

blue

Black-eyed Susan

yellow

Violet

purple

Poppy

orange

Indian Paintbrush

red

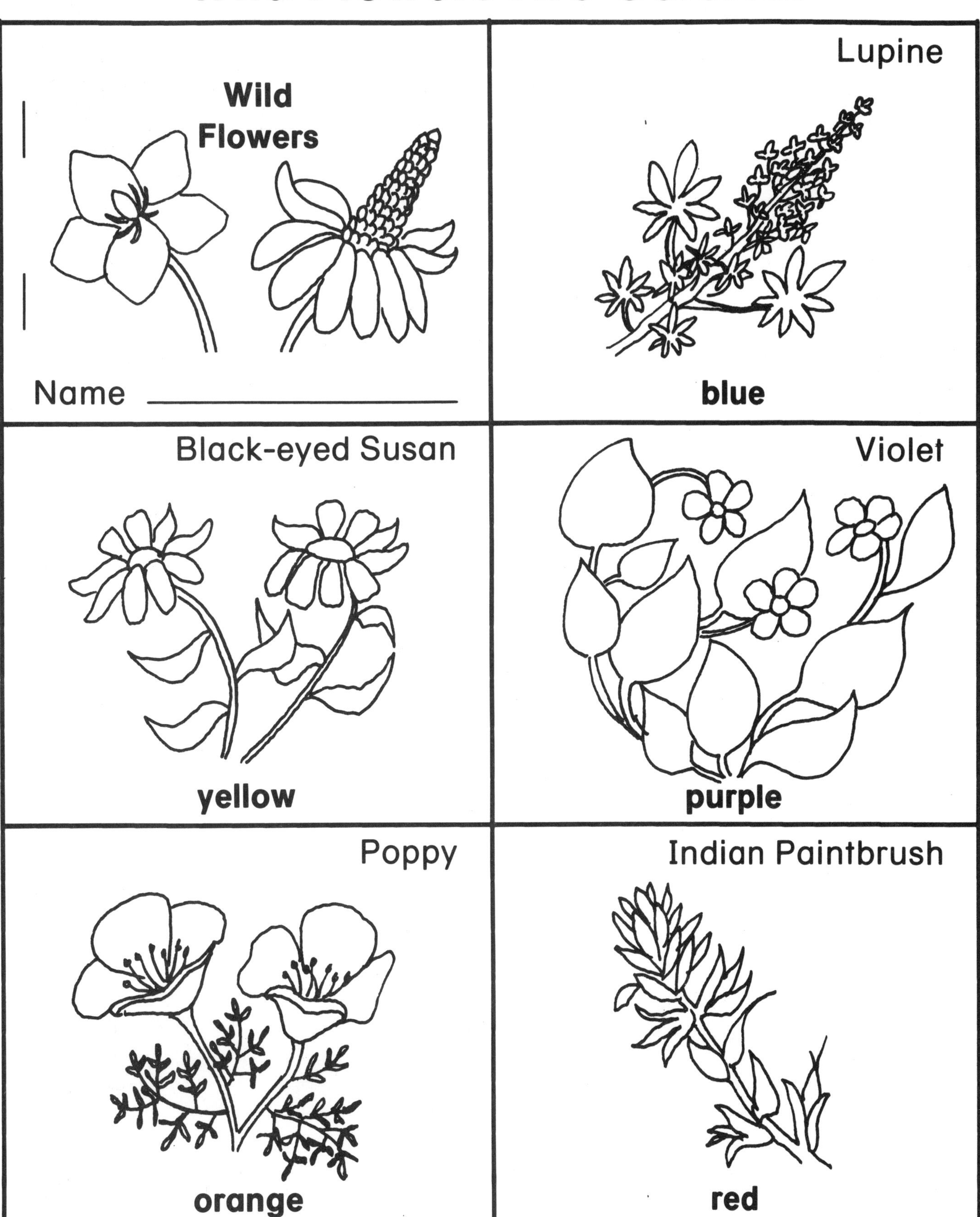

Name ______________________________ Draw a line to connect the matching flowers. Color.

Flowers Are Different Shapes

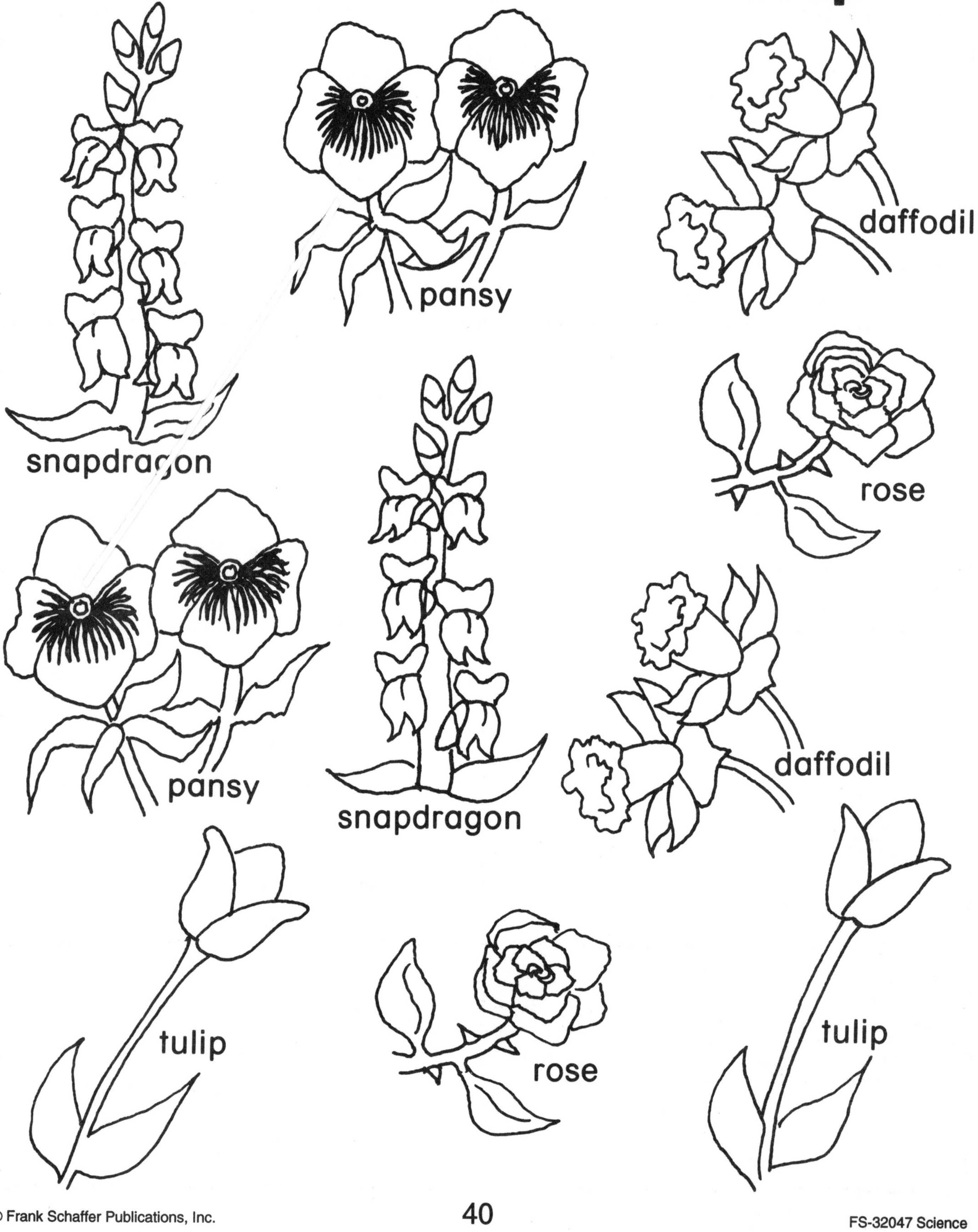

Name ______________________________ Cut and paste in order.

Some Flowers Grow From Bulbs

1	2
Plant a bulb.	Winter comes.
3	4
Spring comes.	It is a flower.

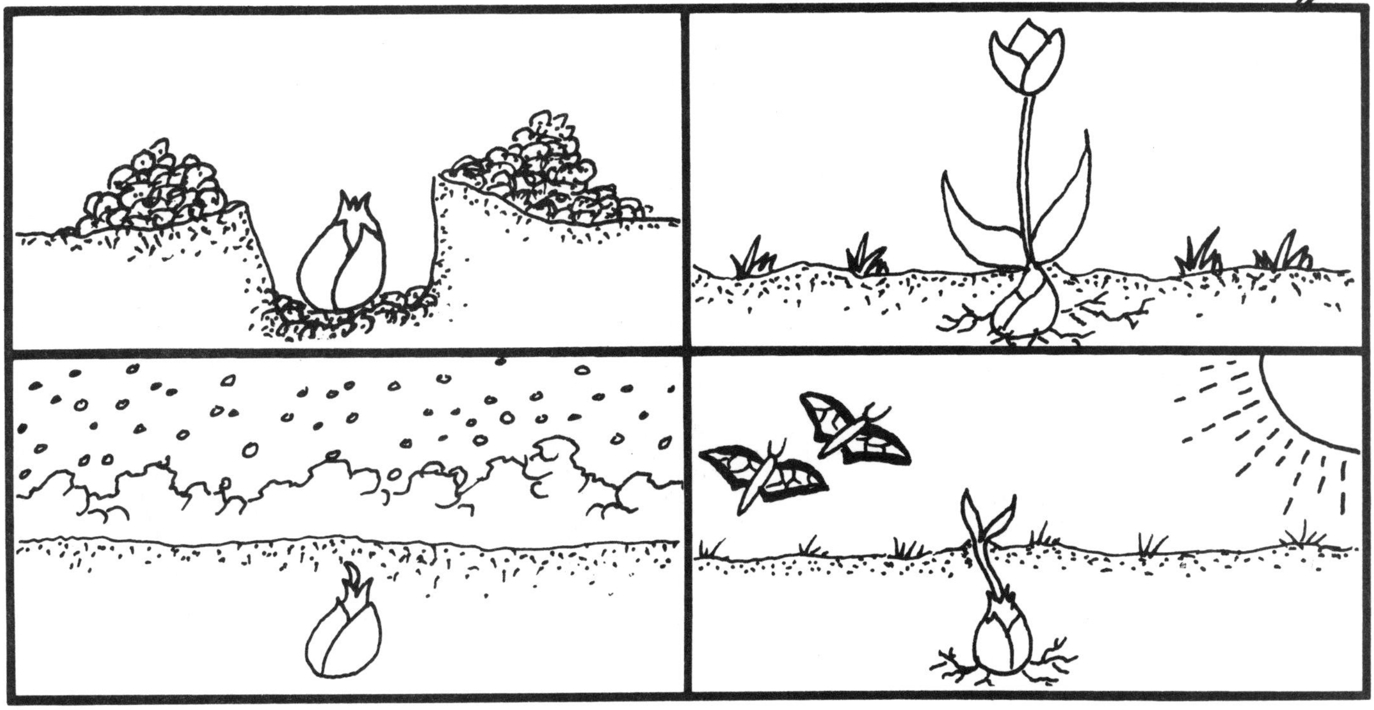

Name ______________________________ Color and remember.

What Do Flowers Give Us?

ANIMALS

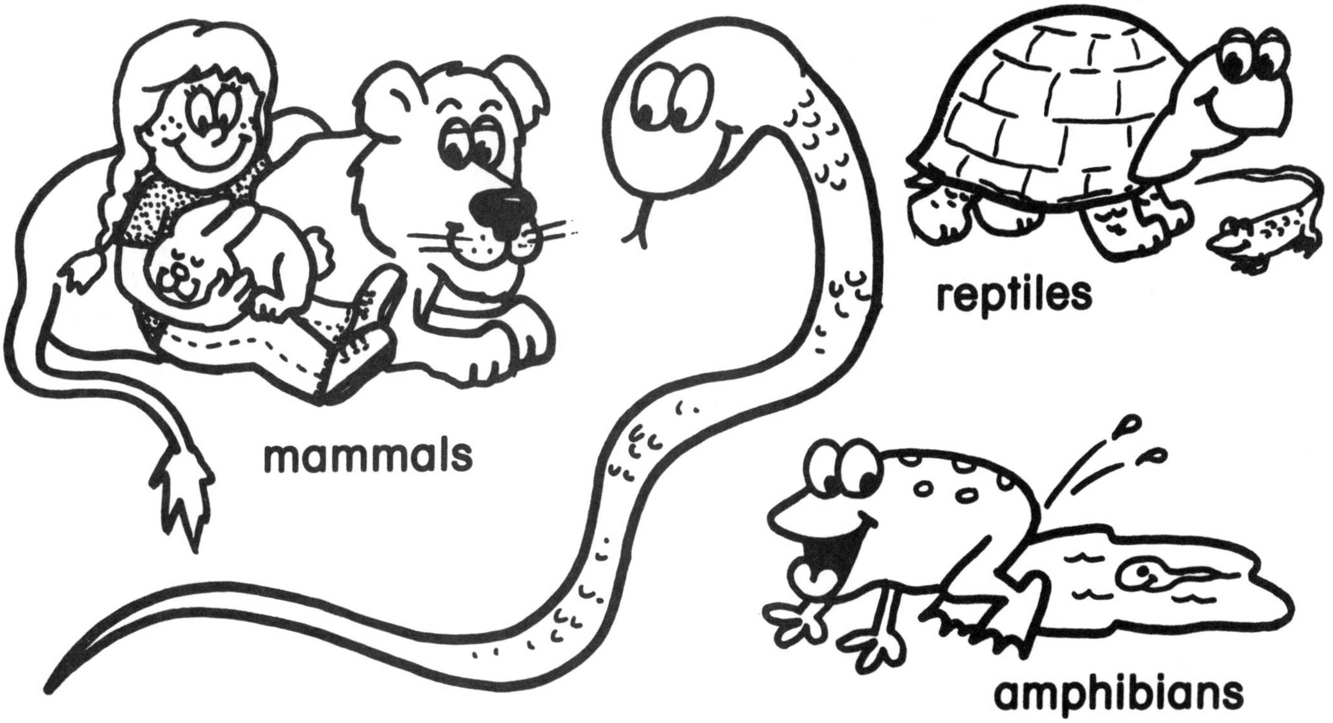

This book belongs to: ____________________

Name ________________________ Color, cut, and paste to complete the reptiles.

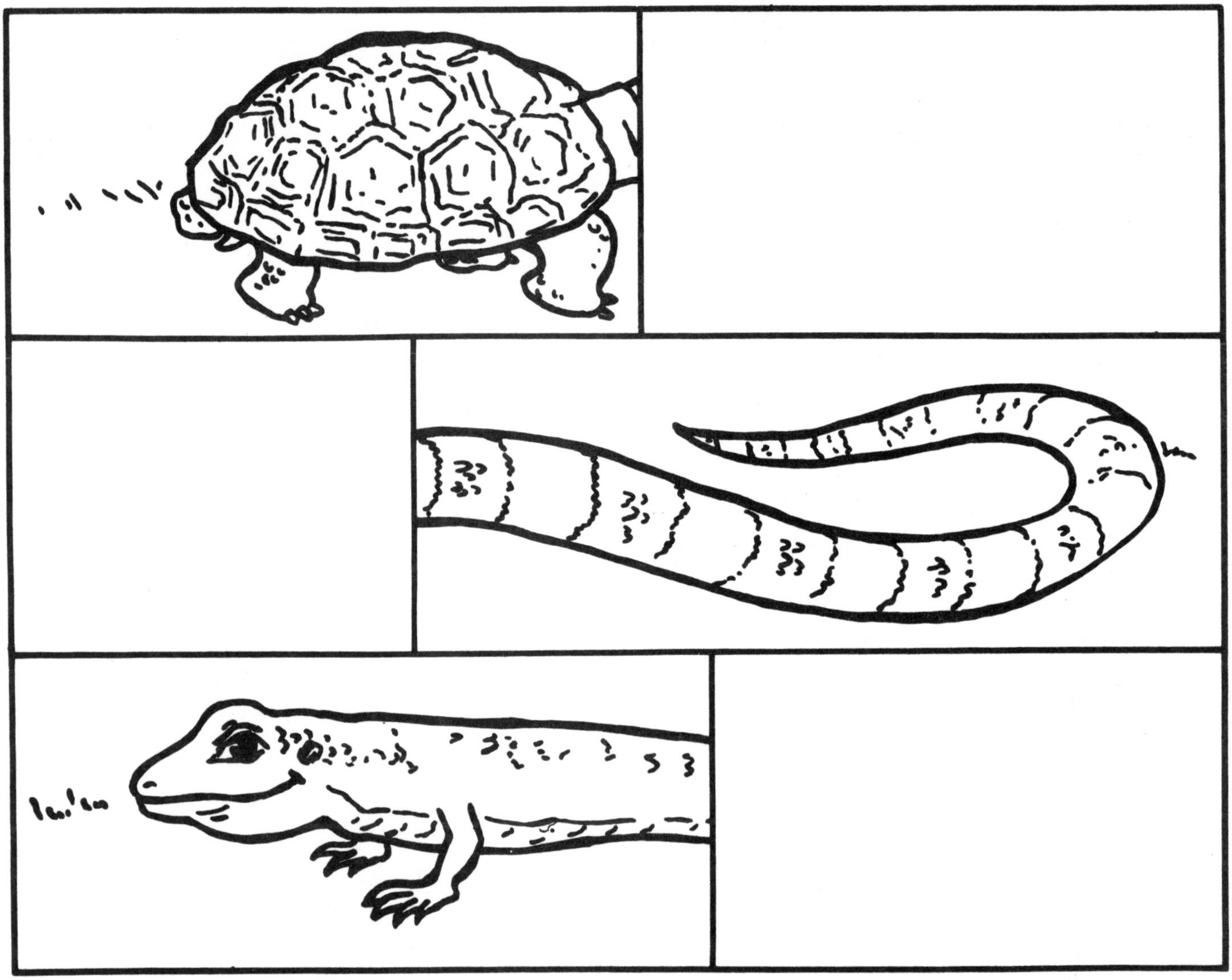

Reptiles have scaly skin.

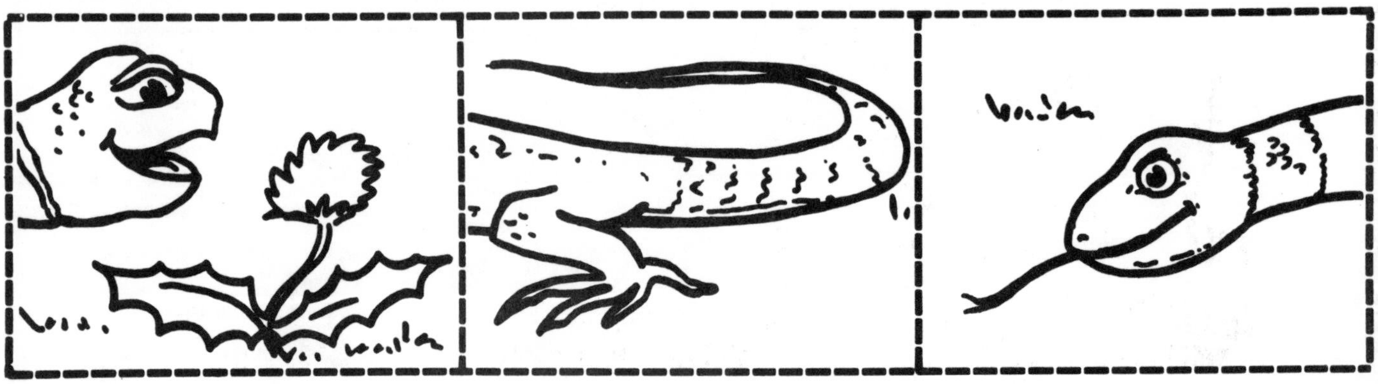

Name ______________________________ Color only the mammals.

Mammals

Name ______________________ Draw a line to connect the matching fish. Color the fish.

FISH
Fish breathe under water.
Fish come in many shapes.

Name ______________________________ Cut out and paste in order.

AMPHIBIANS

1	2
3	4

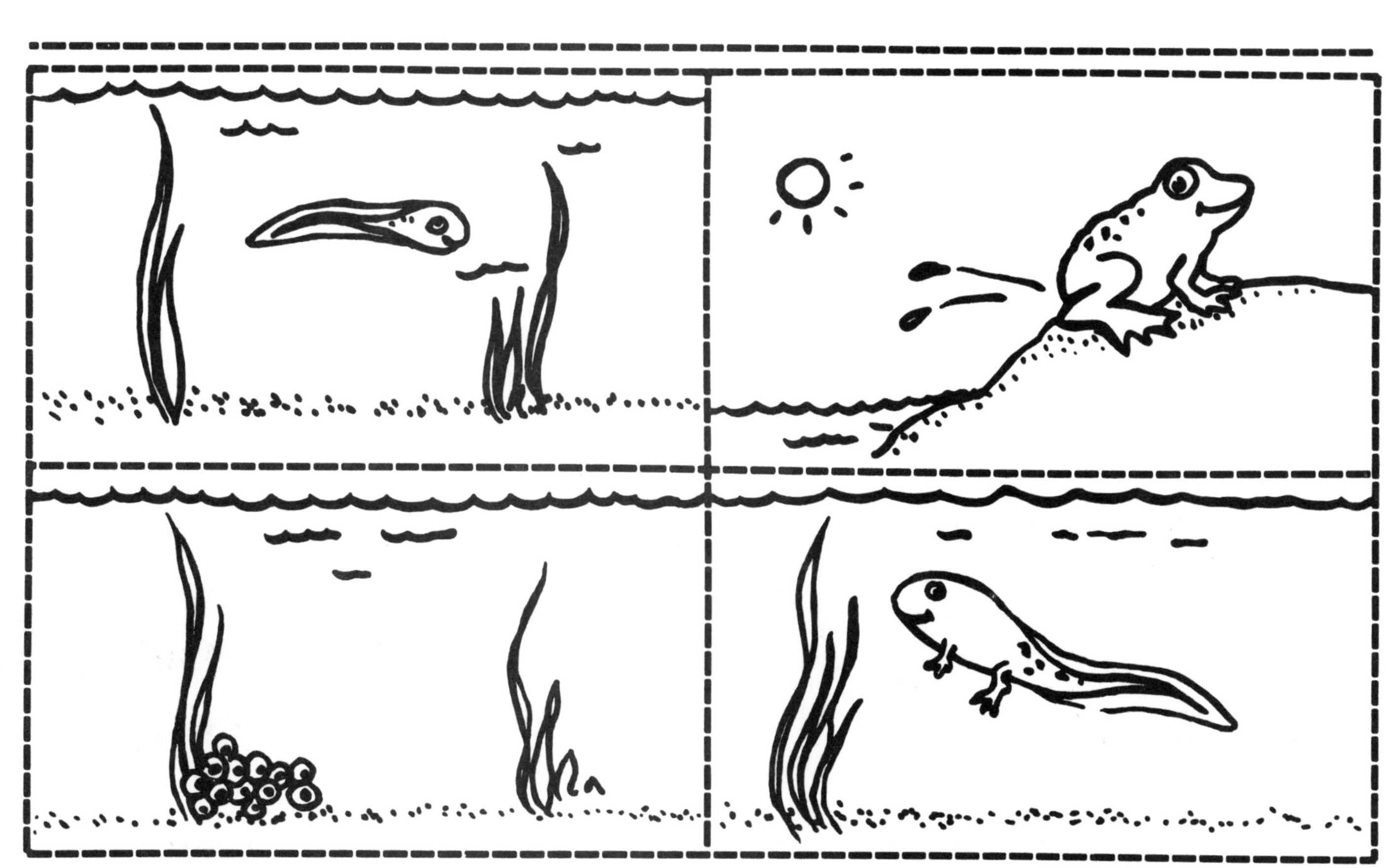

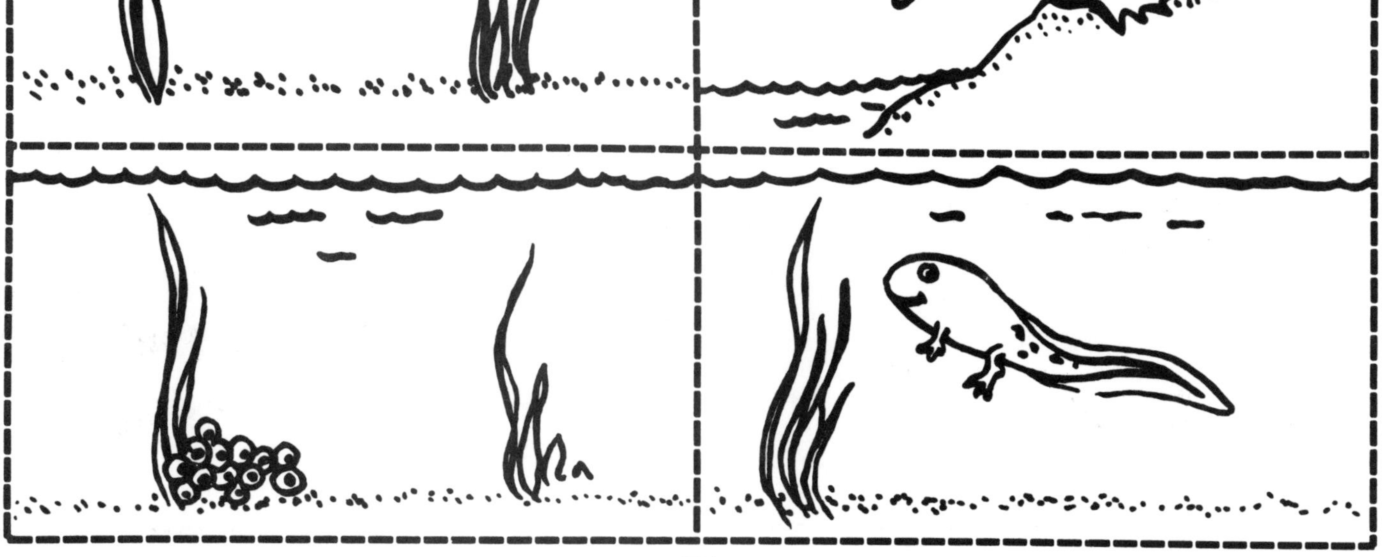

Name ______________________________ Color only the birds.

BIRDS

Birds have feathers.
Many birds fly.
Birds have beaks.
Birds have 2 legs.

Baby Animals

This book belongs to ____________________

Name ______________________________ Color the baby animals that ride. Color their parents.

Some baby animals ride.

Name ______________________ Color, trace and remember.

Mother bear watches her cub.

Brown Bears

cub

Name ______________________________ Cut and paste in order.

A Baby Chick

1	2
A hen lays the egg.	A tiny chick grows inside.
3	4
The chick hatches.	The chick grows bigger.

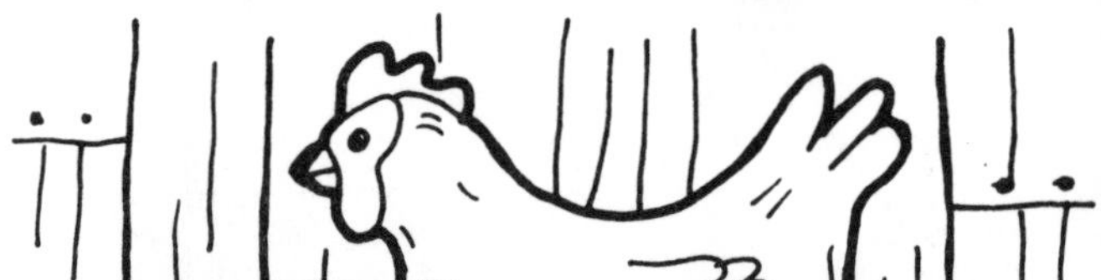

Name ______________________ Read, color and remember.

Little Whale

Father whale sings to his baby.
Little whale hears him.
She swims home.

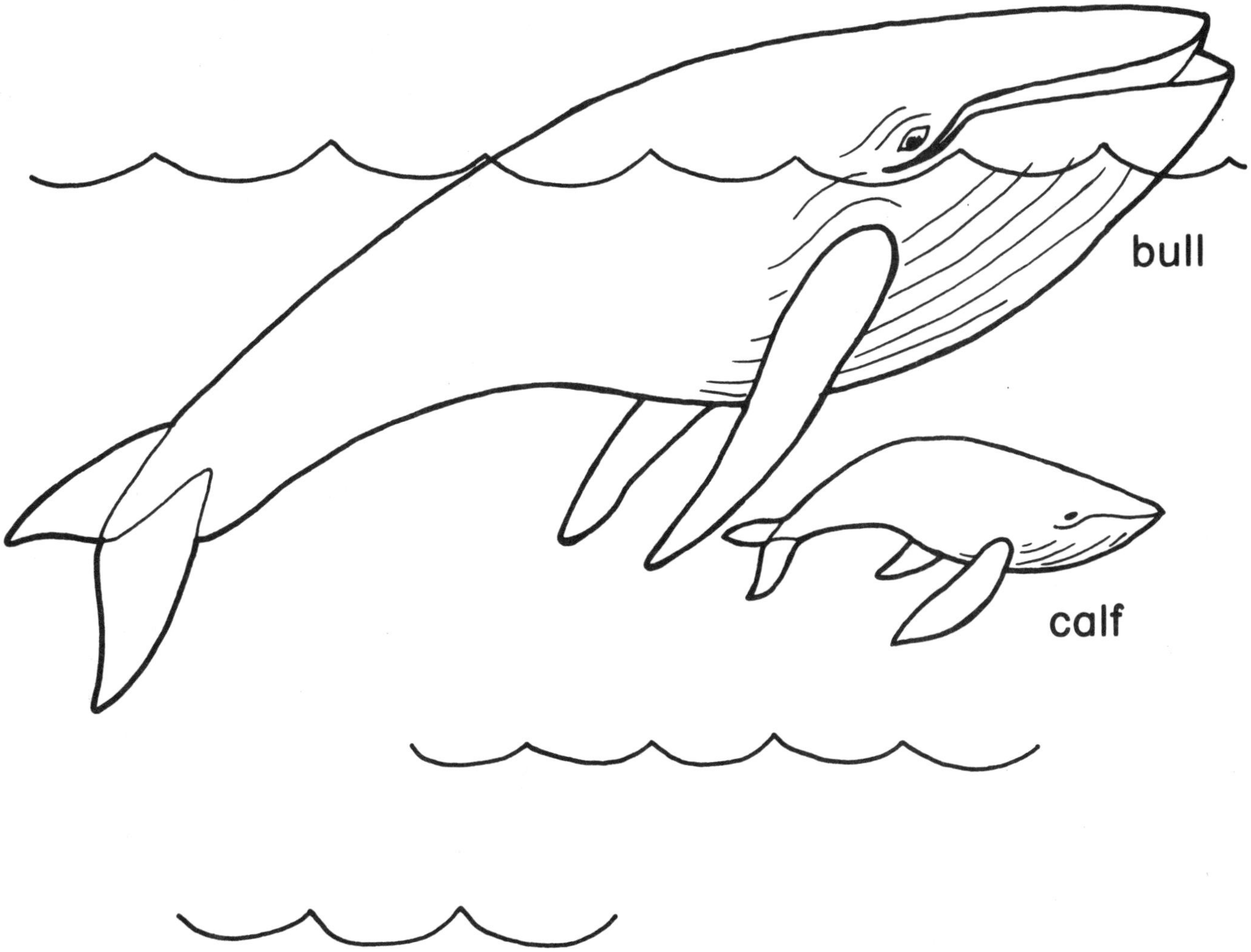

Humpback Whales

Name ________________________________ Read, color, remember.

Color the fawn brown with white spots.
Color the forest brown and green.

The colors of the fawn help it hide.

Name ______________________________ Read, count and color.

A Litter

These babies were born at the same time.
They are called a litter.
How many babies are in this litter? __________

Hamsters

Endangered Animals

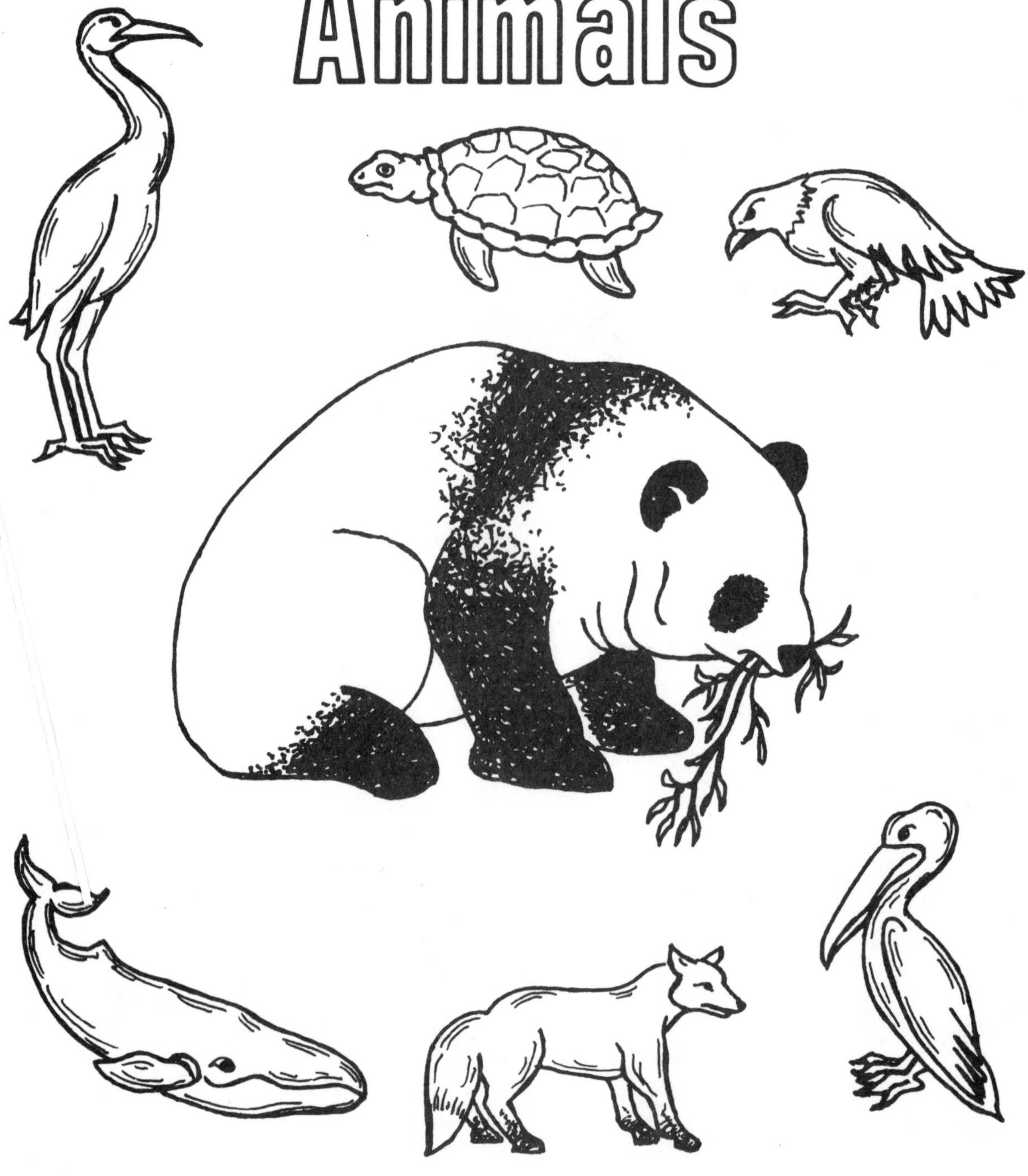

This book belongs to ______________________

Name ______________________________ Match the animals. Color.

Some Endangered Animals

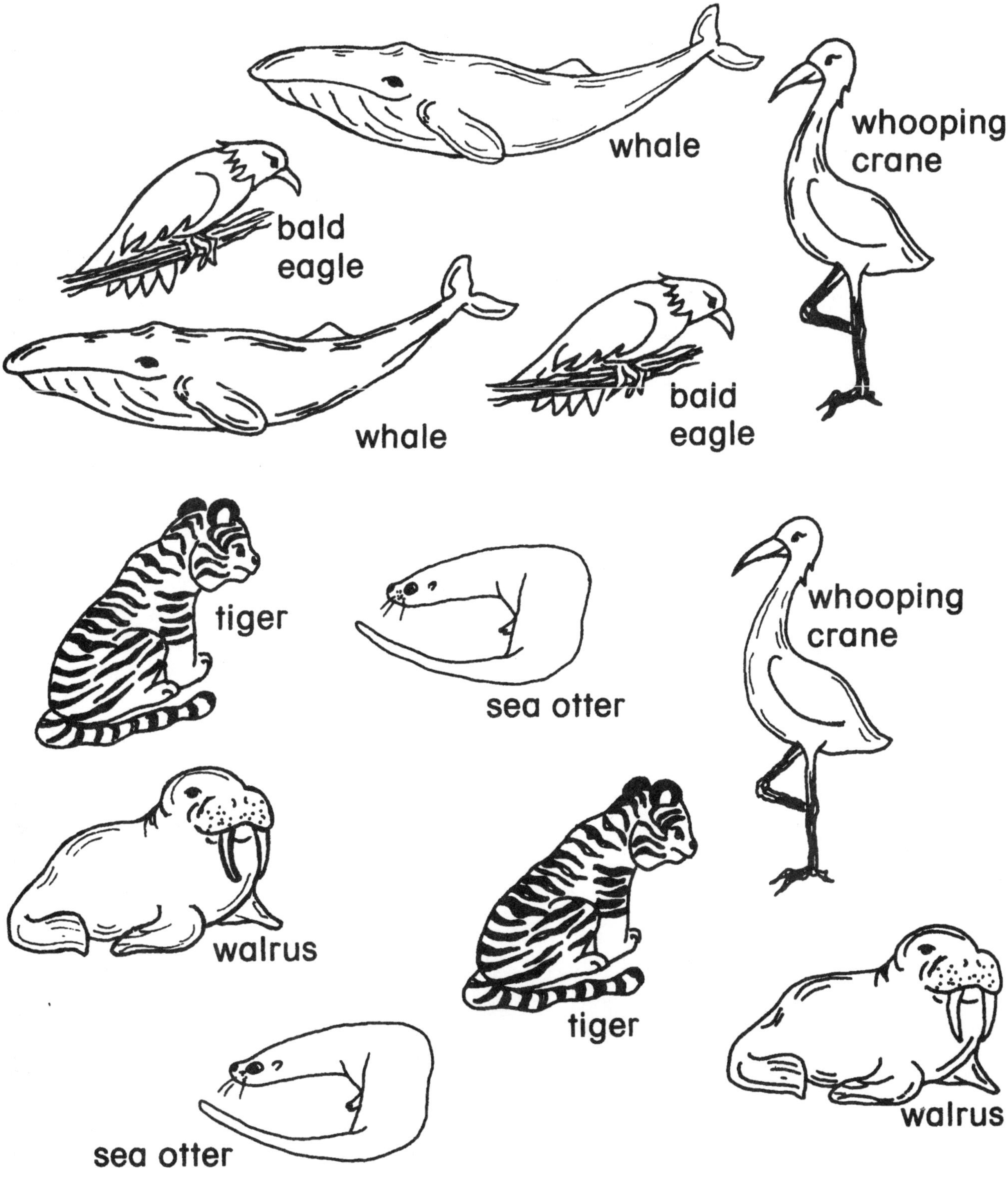

Name ______________________ Color and remember.

People made laws to protect the koala.

Name ______________________________ Trace the dotted lines and color.

Help these animals find their homes.

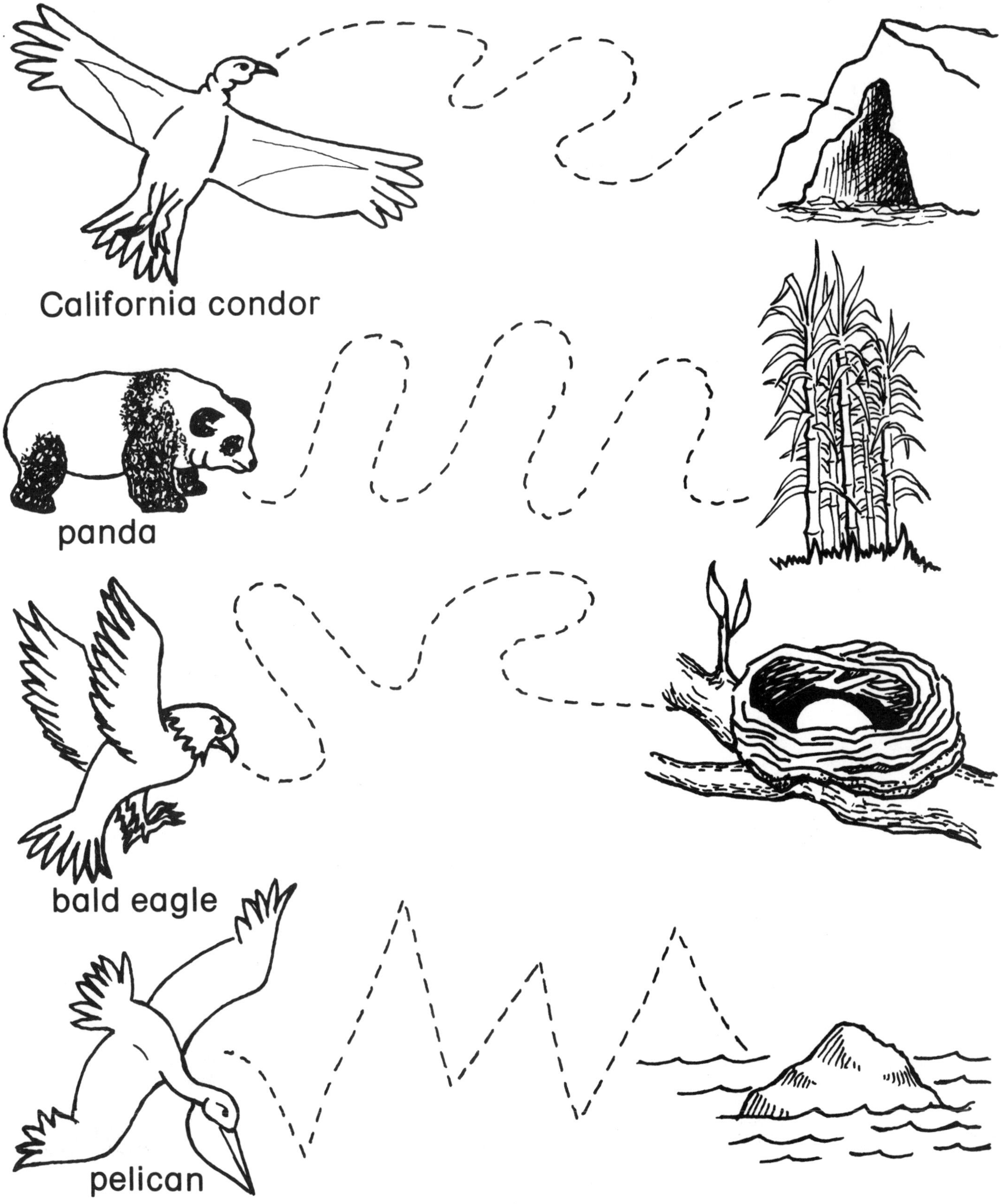

Name ______________________________

Cut and paste in the correct place.

Endangered Mammals		Endangered Birds	

Name ______________________________ Trace and write. Color.

Bison live in Yellowstone National Park.

bison

Name ______________________________ Read and color.

More Endangered Animals

brown

bear

black

gorilla

black

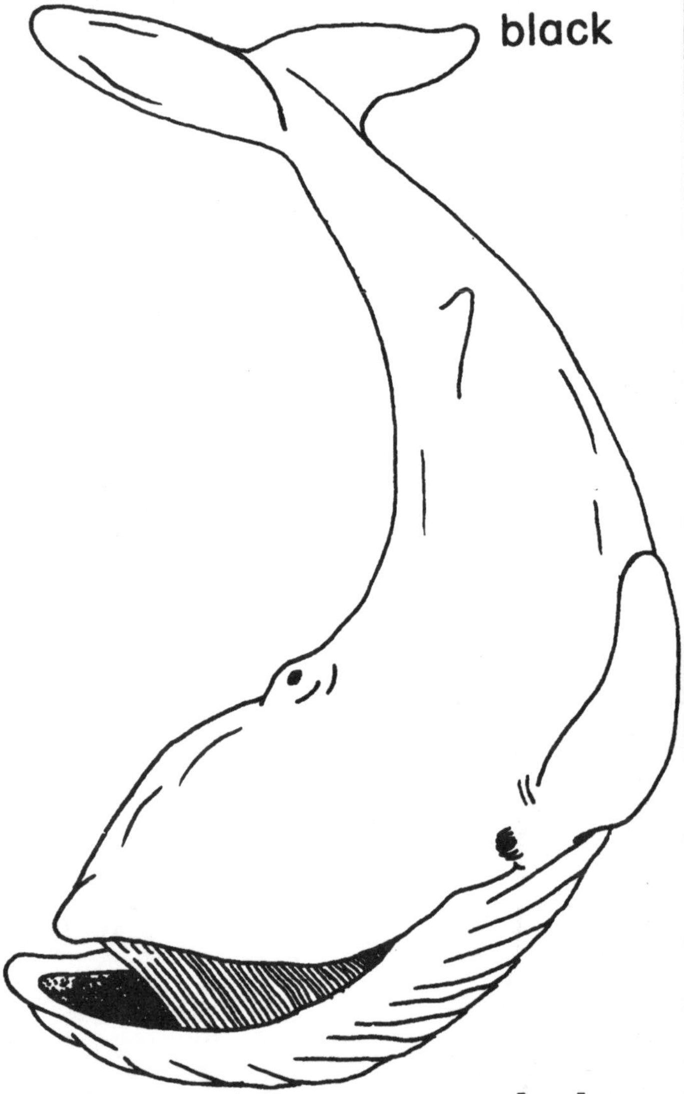

whale

red

wolf

green

turtle

Dinosaurs

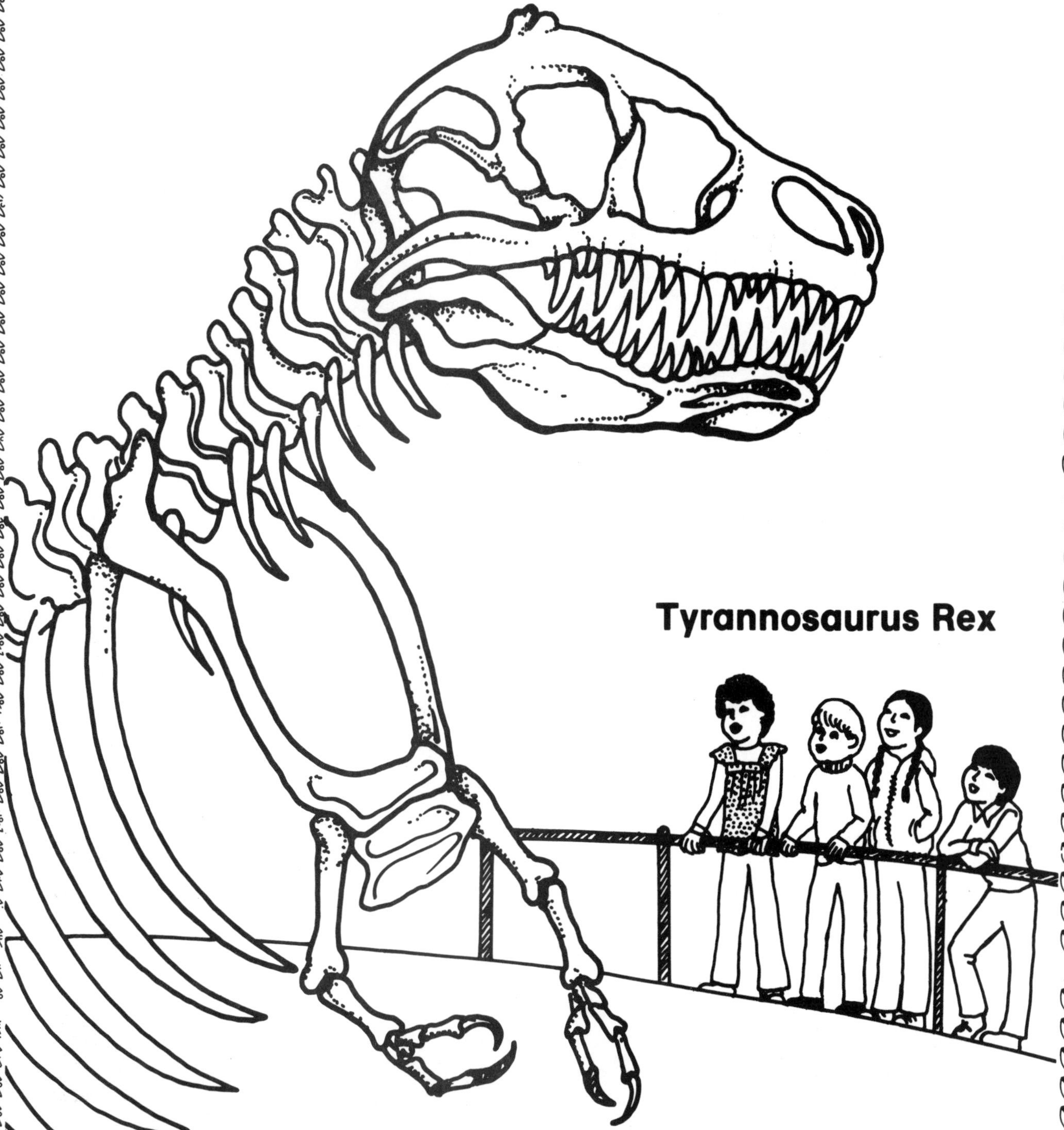

This book belongs to ______________________________

Name ______________________________ Color only the dinosaurs.

Dinosaurs lived millions of years ago.

Name ______________________ Color, cut and staple to make a book.

The Discovery of Supersaurus

Name ______________________

1

Dinosaur Jim dug for dinosaur bones.

2

One day he found a gigantic bone.

3

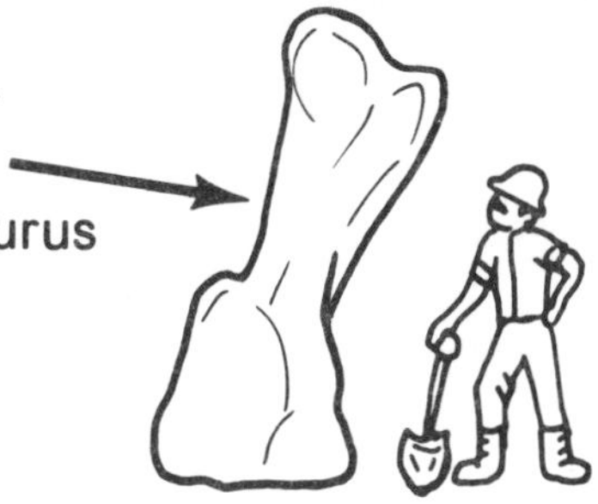

He found more and more huge bones.

4

He had discovered an enormous new dinosaur.

5

Dinosaur Jim named the new dinosaur **SUPERSAURUS**.

6

Supersaurus weighed 70 tons, as much as 14 elephants.

7

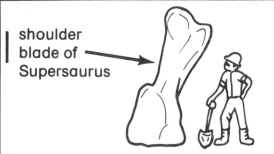

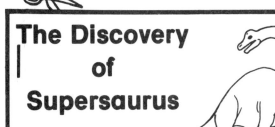

Dinosaur Jim has 16 tons of bones to sort!

Name ________________________________ Read. Cut and paste the words to match the pictures.

Where did dinosaurs live?

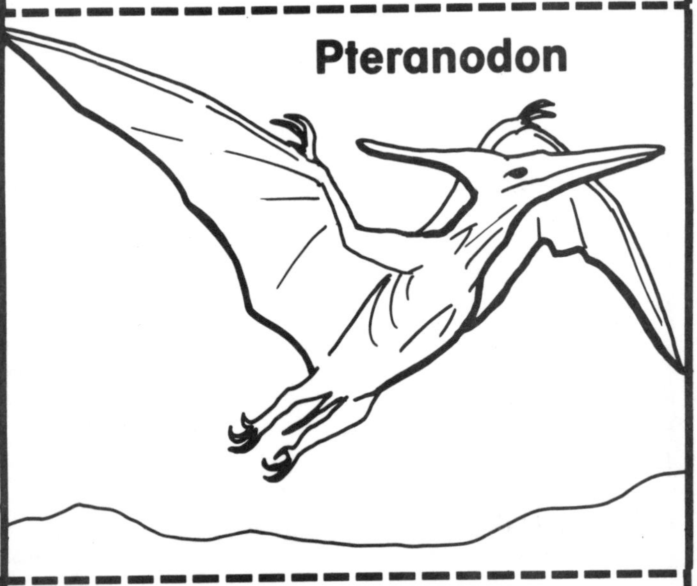

in the water	in the forest
in the air	near the water

Name ______________________________ Cut and paste in order.

Scientists Dig for Dinosaur Bones

1	2
3	4

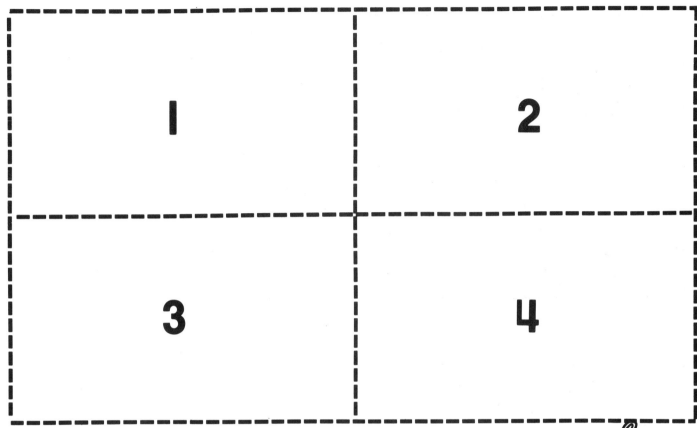

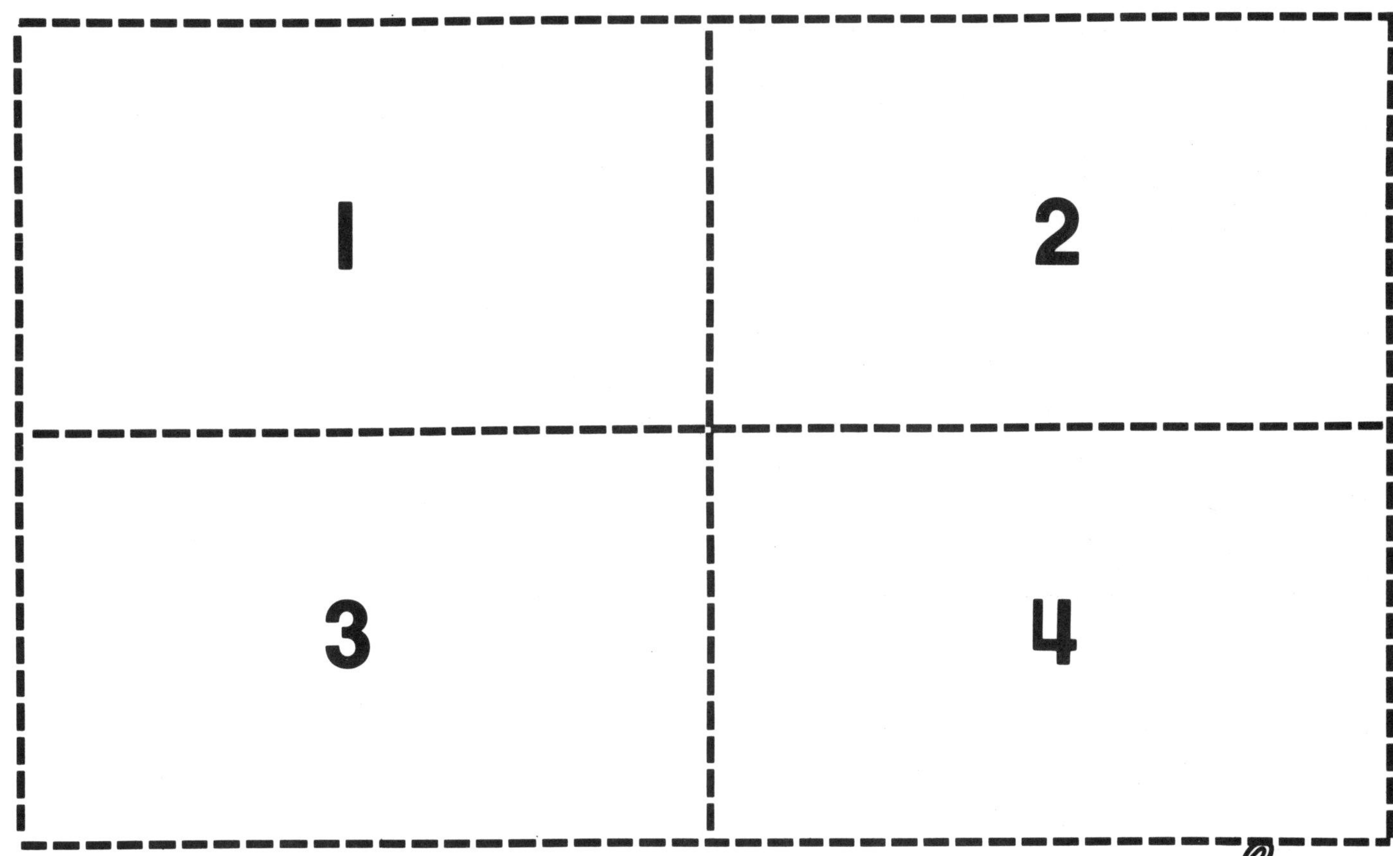

Name ______________________________ Cut and paste in order.

Tyrannosaurus Rex is the fiercest animal that ever lived.

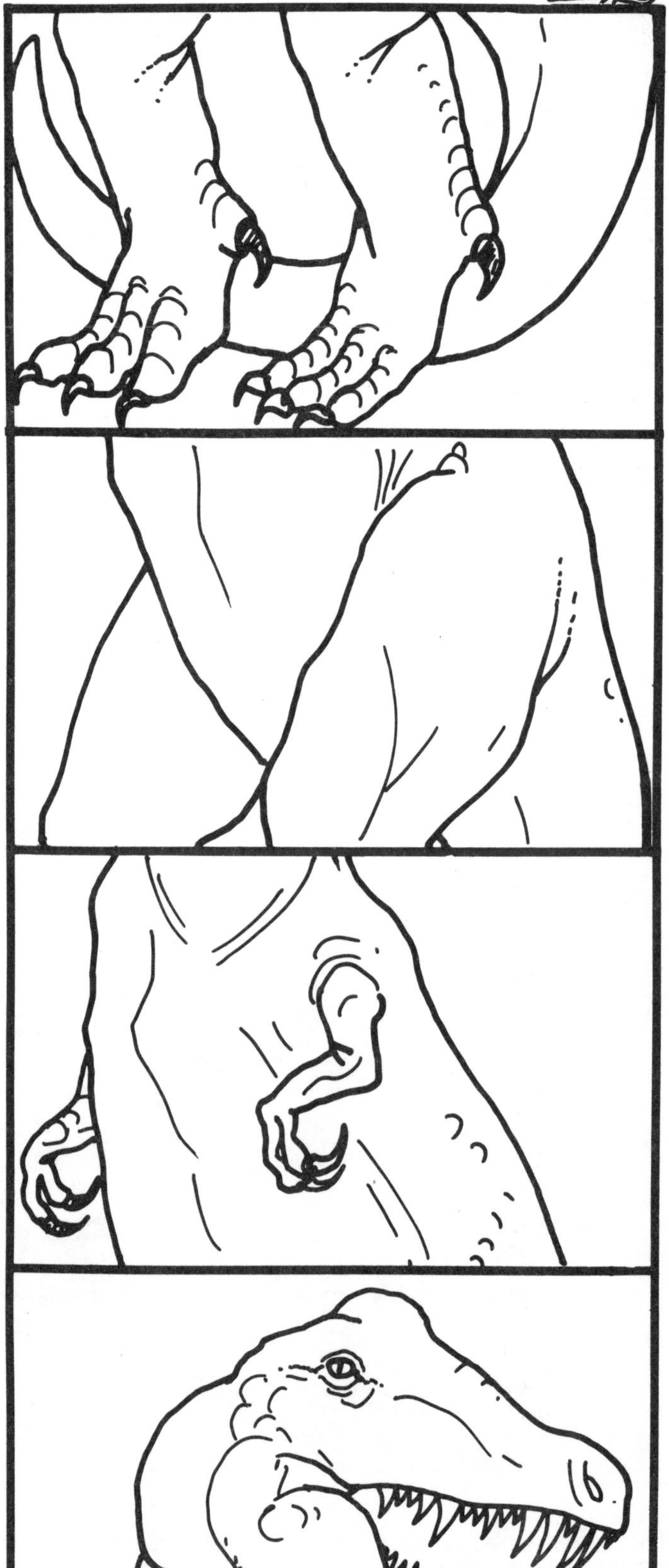

Name ________________________ Cut and paste to measure the dinosaur.

Diplodocus was one of the longest dinosaurs.

1	2	3

Diplodocus was as long as three school buses.

Ecology

This book belongs to ______________________________

Name ____________________ Color and remember.

Ecology is the study of plants, animals, air, water and the way people treat them.

ecology

Name ______________________ Cut and paste the labels.

Environment includes many parts.

5. **people**	1. **plants**	2. **air**
4. **water**	6. **soil**	3. **animals**

Name ______________________________ Cut and paste to match the pictures.

Where Do You Find Pollution?

in the street	in the air
by the road	in the water

Name ______________________ Cut and fold back.
Turn over.

Help Clean Up Our World

fold back fold back

Name ______________________________

Trace the dotted lines and color.

I Can Help Clean Up the World!

My Book About

Seasons

Name

Name ______________________________

The Sun

Sunshine helps make the seasons.

Color, cut, and paste.

Name

Summer

It is hot in the summer.

Name ____________________

Fall

Fall days are cool.
Leaves fall from the trees.

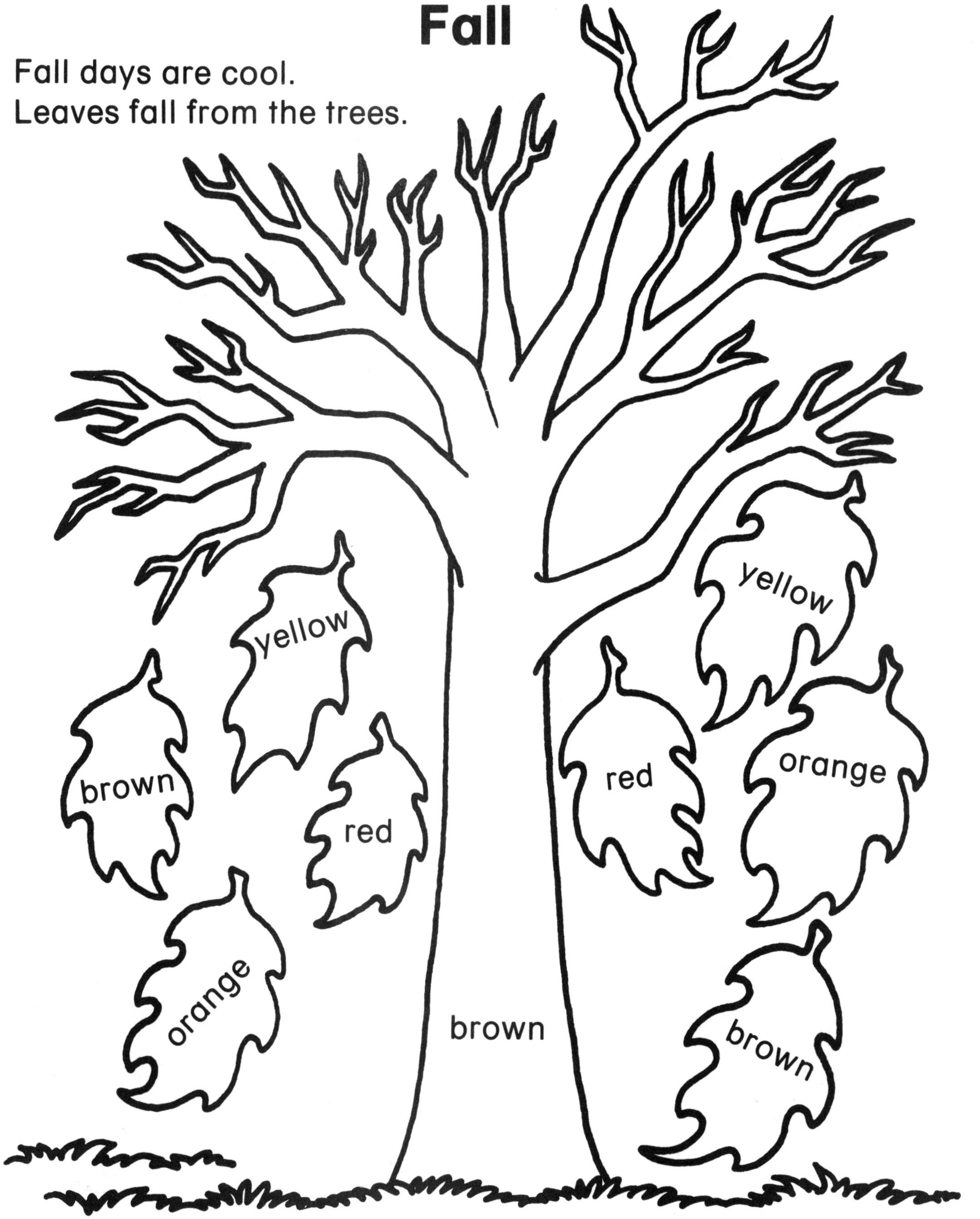

Trace and color.

Name

Winter

Winter is cold.

Name

Spring

Spring is warm.
New plants grow.

Name ______________________

Clothes for the Seasons

winter

spring

summer

fall

Name ______________________

The Four Seasons

Trace, color, and cut to make a puzzle.

Following directions

Four Seasons Flipper

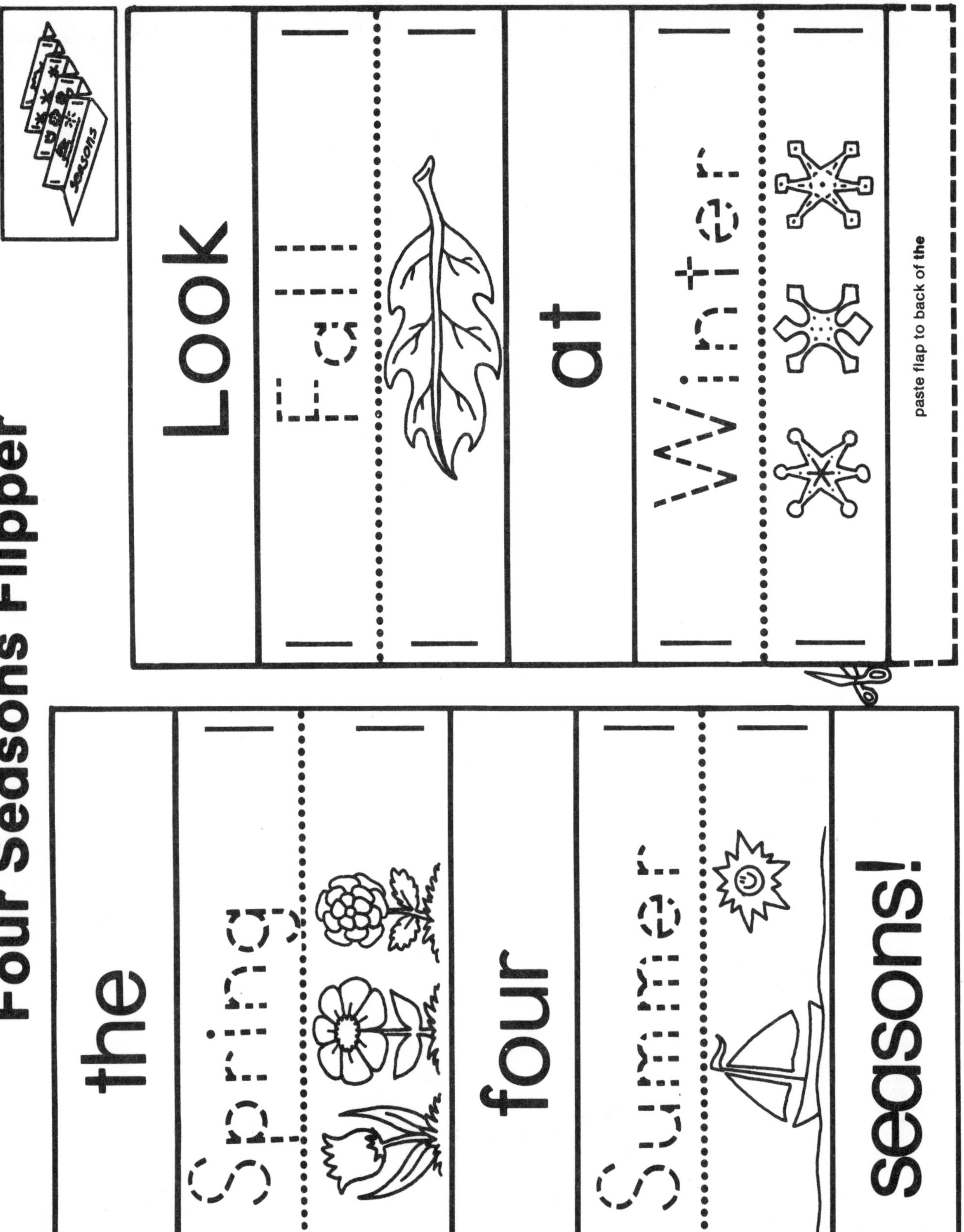

Teacher: Have children do the following 1) Trace the words and color the pictures. 2) Cut out the pieces. 3) Glue where indicated. Help children fold on the dotted lines and staple where shown to form four flaps. Direct children to lay the project flat on a table, and fold the flaps back and forth on the solid lines to make them stand. When children look directly at the project they will see a message. Then they can flatten the flippers downward to see the names of the seasons and upward to see seasonal pictures.

My Book About

Weather

Name

Name______________________________

Weather Words

Write a weather word in each set of boxes.
Color the pictures.

sun rain snow wind

Name ______________________

Clouds

Read the poem.
Glue cotton on the cloud shapes.
Color the picture.

I like to see
What clouds can be,
As they go by
Up in the sky.

Name

Rain Falls From Clouds

Trace the words.
Color the picture.
Make the raindrops.

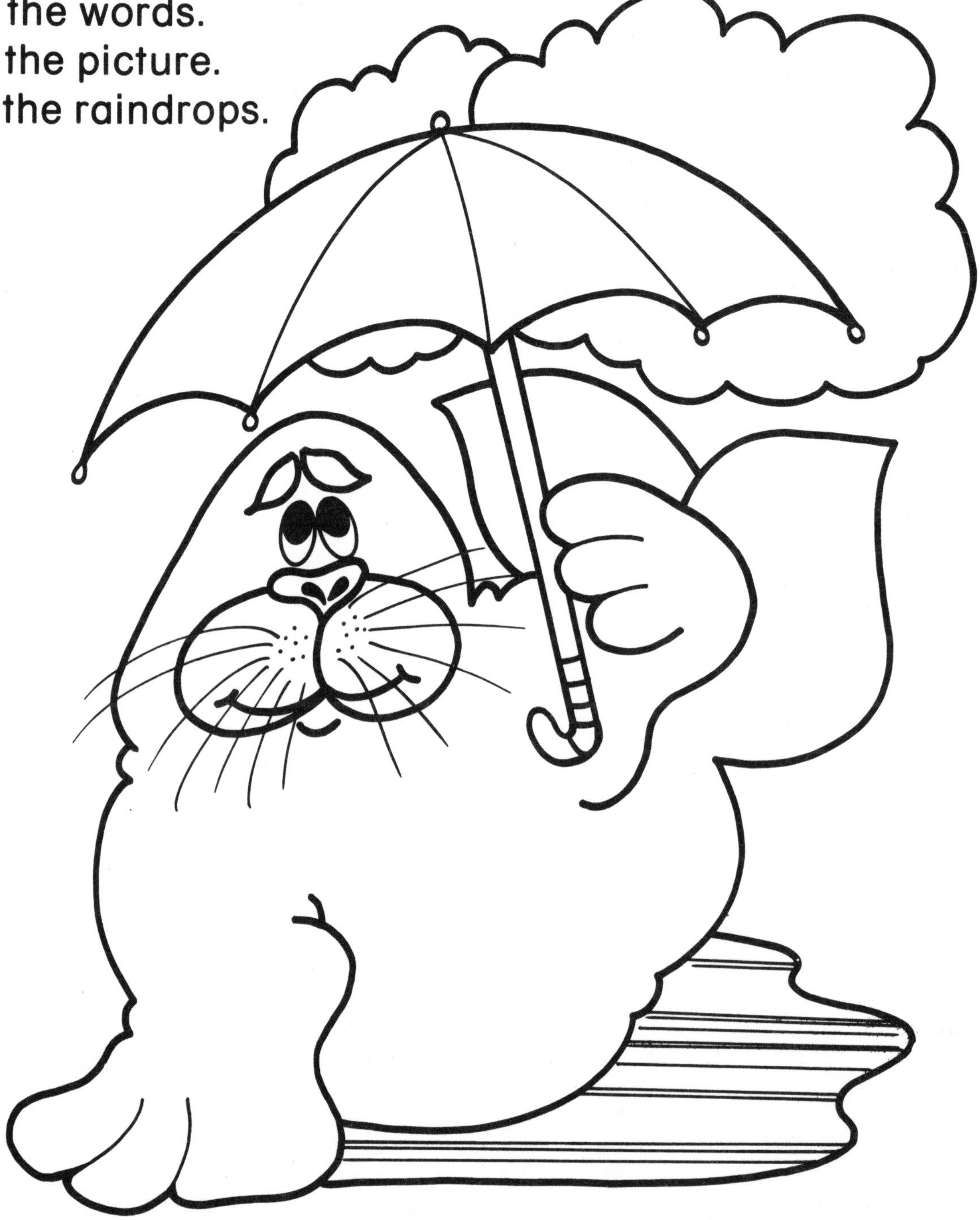

Rain falls from clouds.

Teacher: Have students dip a fingertip into blue paint and press it onto the picture to make raindrops.

Name ____________________

Snow Is Flakes of Ice

Cut, match, and paste.

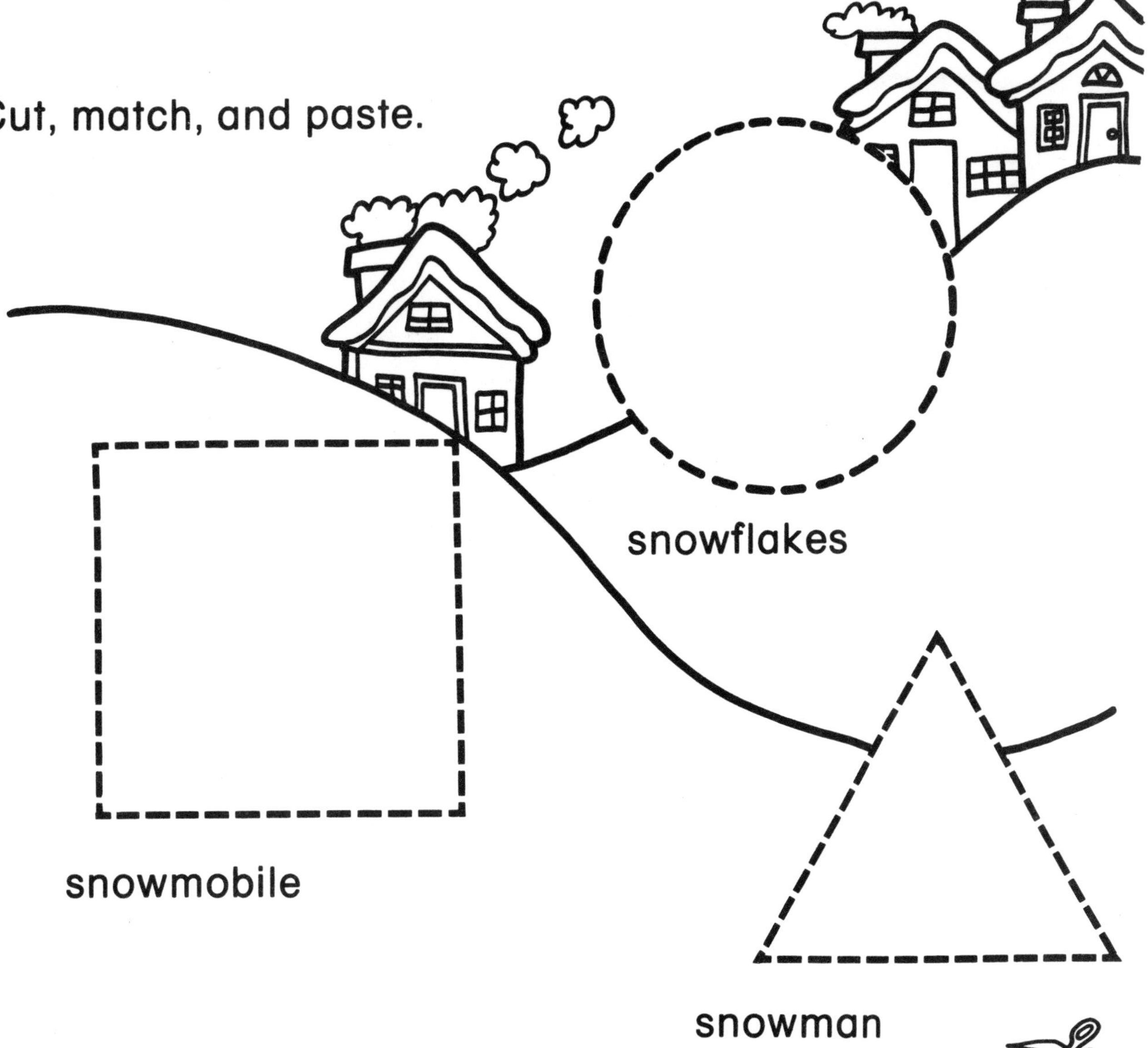

Name ______________________

The Faces of Wind

Cut and paste.

A friend

Not a friend

Name____________________

Sunshine and Rain Make a Rainbow

Trace the word.
Color, cut, and match the puzzle.

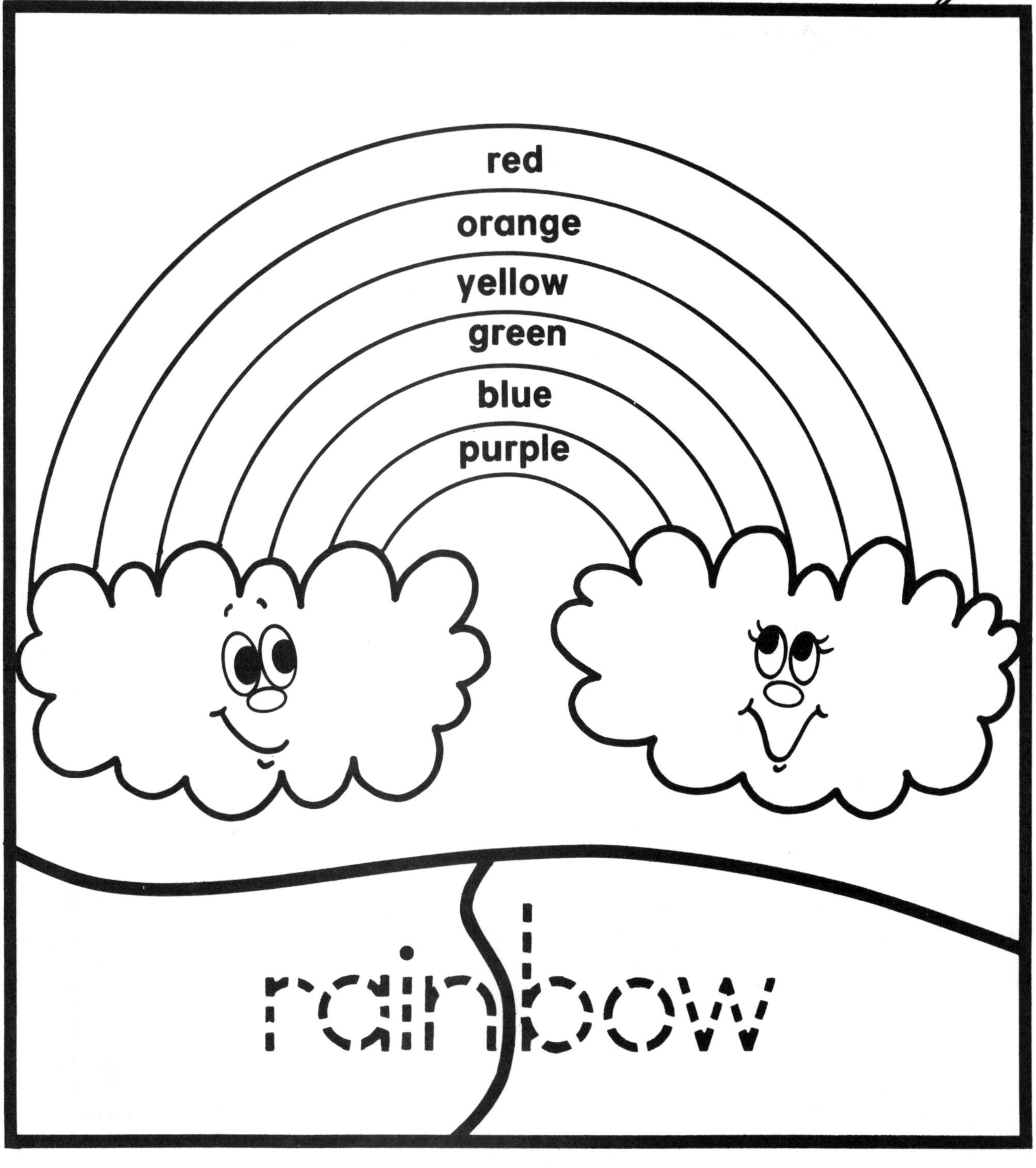

Name____________________

My Weather Chart

Make a weather chart for one week.
Cut and paste pictures in each box to show the weather on that day.

Monday	Tuesday	Wednesday	Thursday	Friday

Teacher: Students cut and paste weather pictures from page 93 on the chart.

Weather Pictures

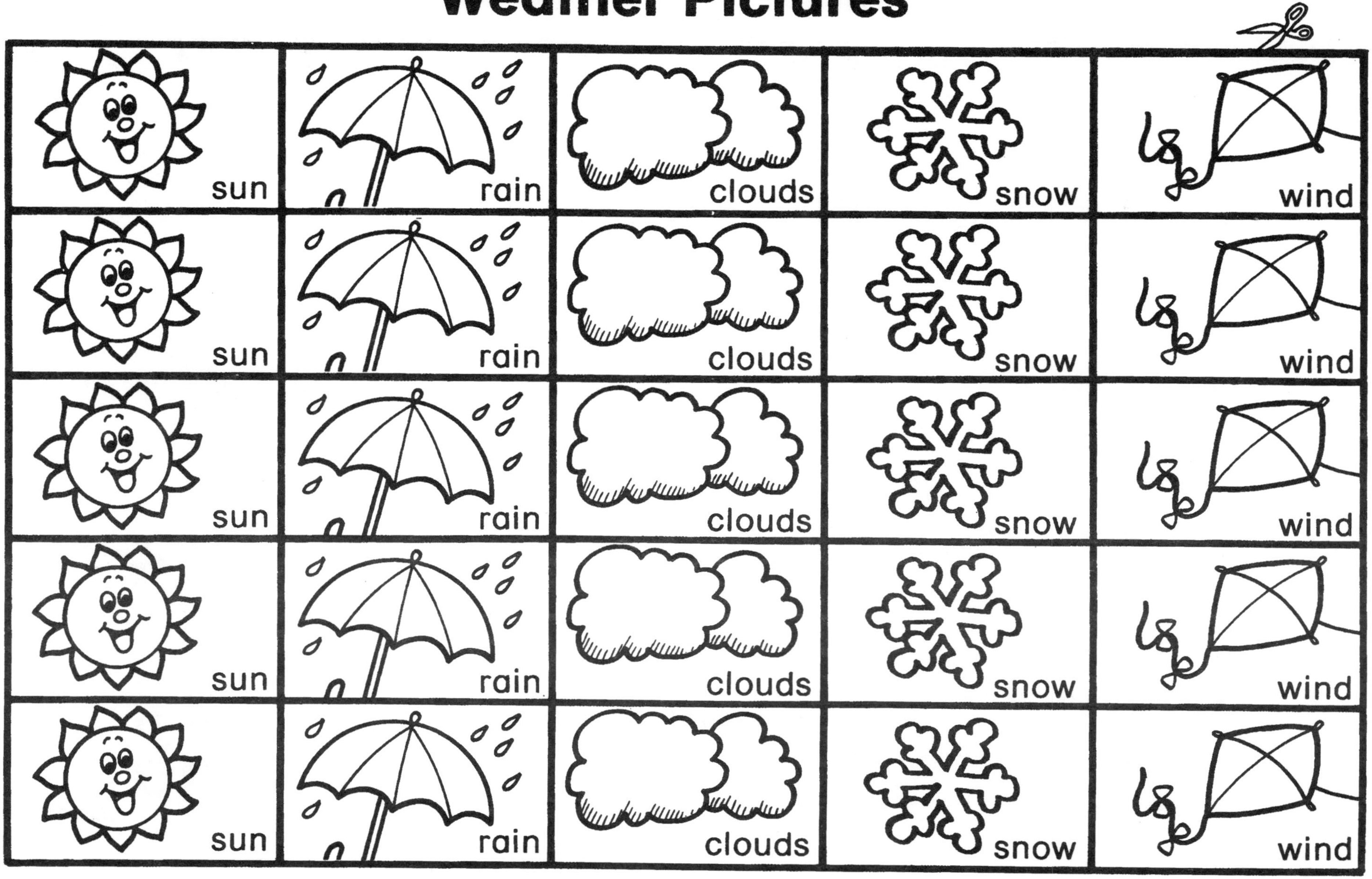

Teacher: Students cut out weather pictures and paste them to the weather chart on page 92.

Name______________________________

Go inside if you see . . . lightning!

Trace and color.
Cut and staple to make a book.

What Is Weather?

Name

Weather is what it is like outside.

1

sunny

2

cloudy

rainy

3

windy

4

snowy

5

SPACE

This book belongs to: ______________________________

Name ______________________ Help Susie get home to Earth.
Follow the maze and color.

We live on planet **Earth.**

Name ______________________________ Color, trace, remember.

One moon orbits the Earth.

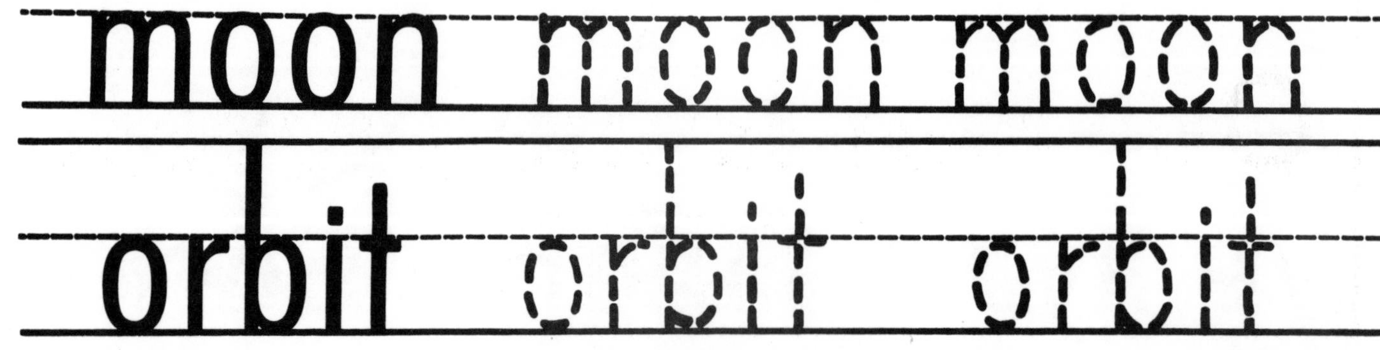

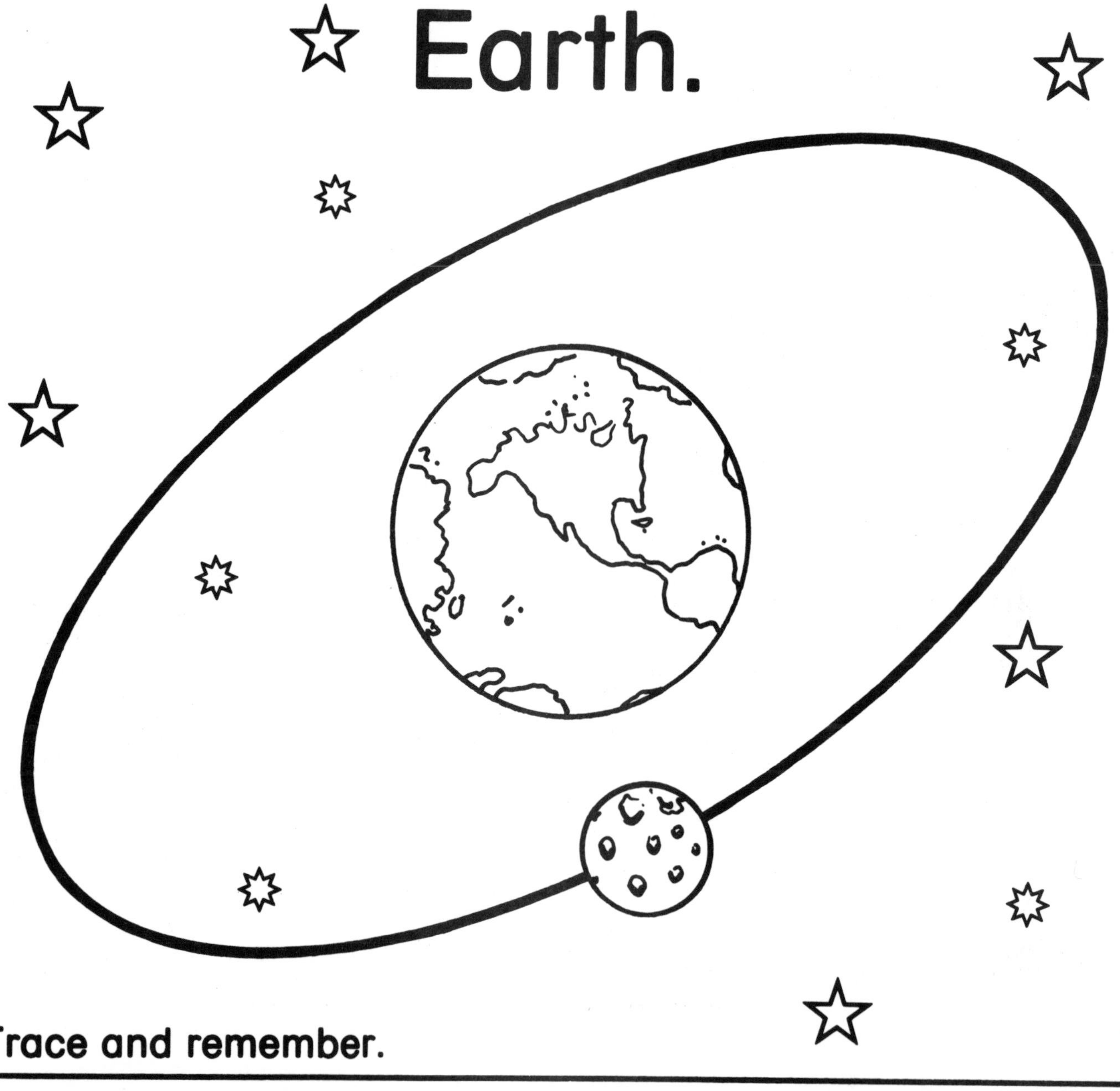

Trace and remember.

moon moon moon

orbit orbit orbit

Name ______________________ Color and remember.

Our sun is the closest STAR.

Name ______________________________ Cut and paste planets on correct orbits.

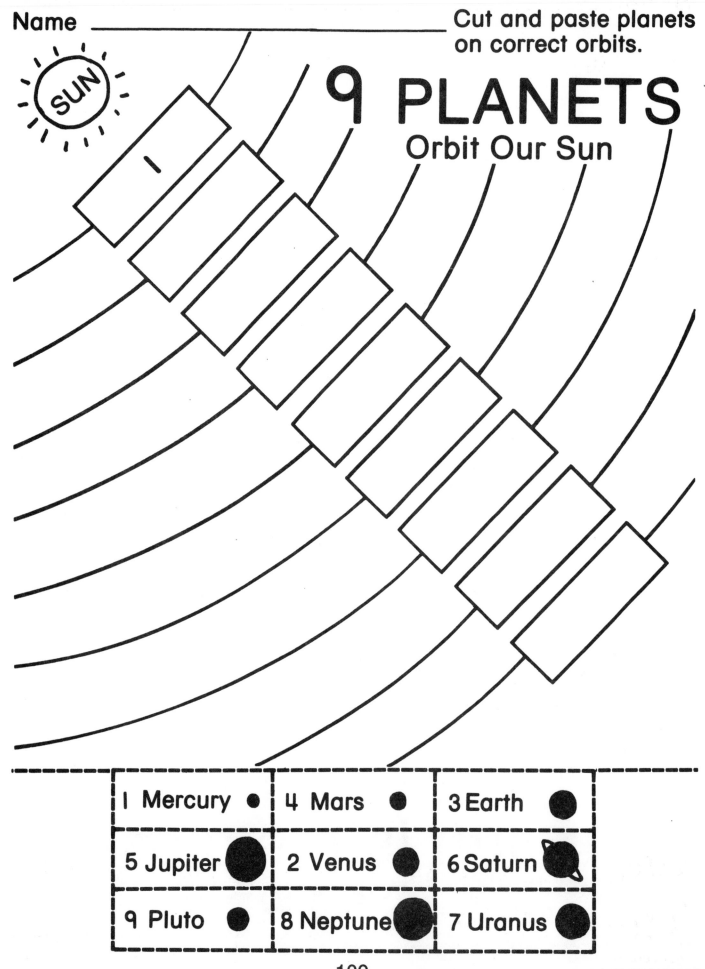

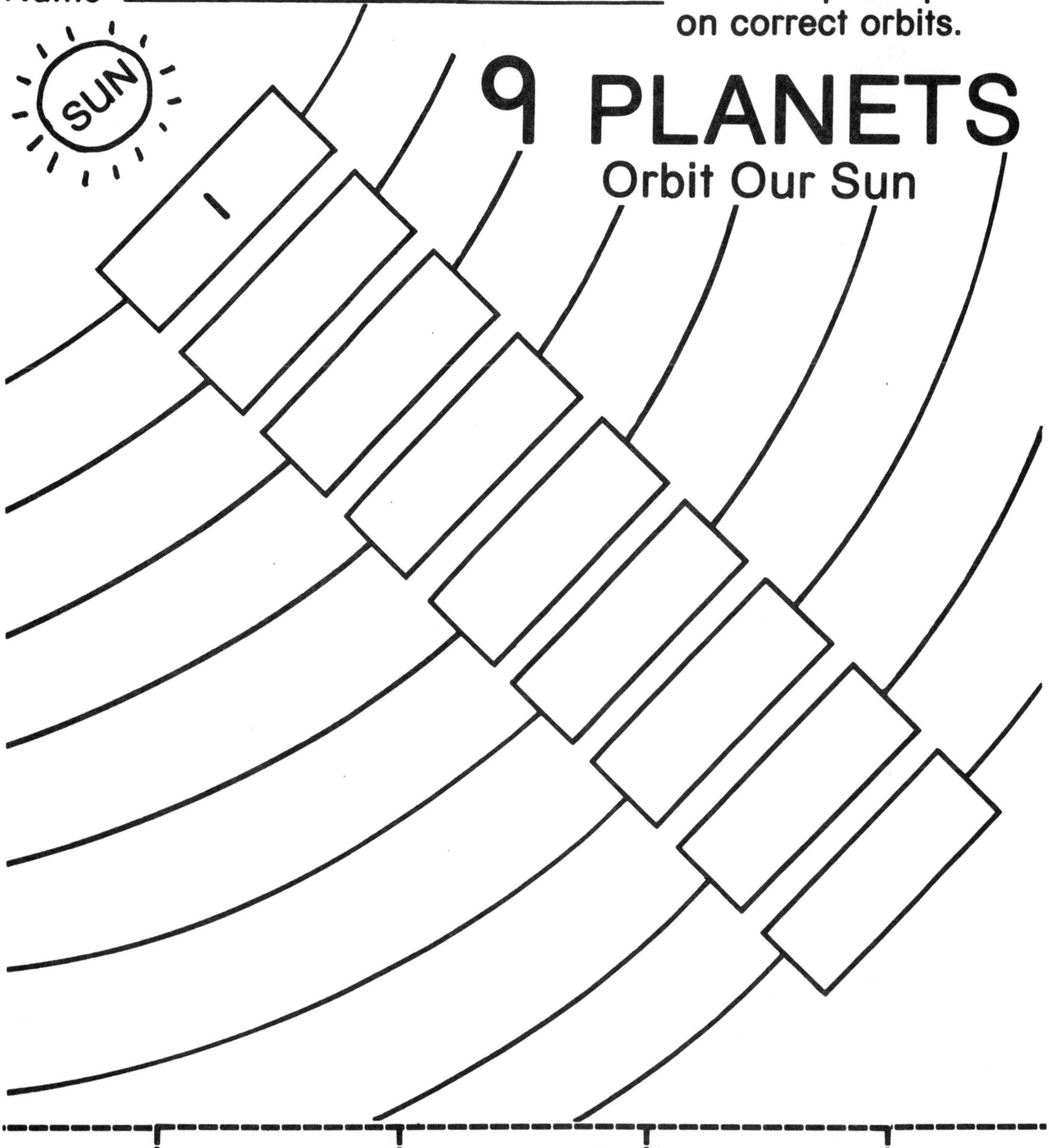

9 PLANETS

Orbit Our Sun

1 Mercury	4 Mars	3 Earth
5 Jupiter	2 Venus	6 Saturn
9 Pluto	8 Neptune	7 Uranus

Name ______________________________ Connect the dots.

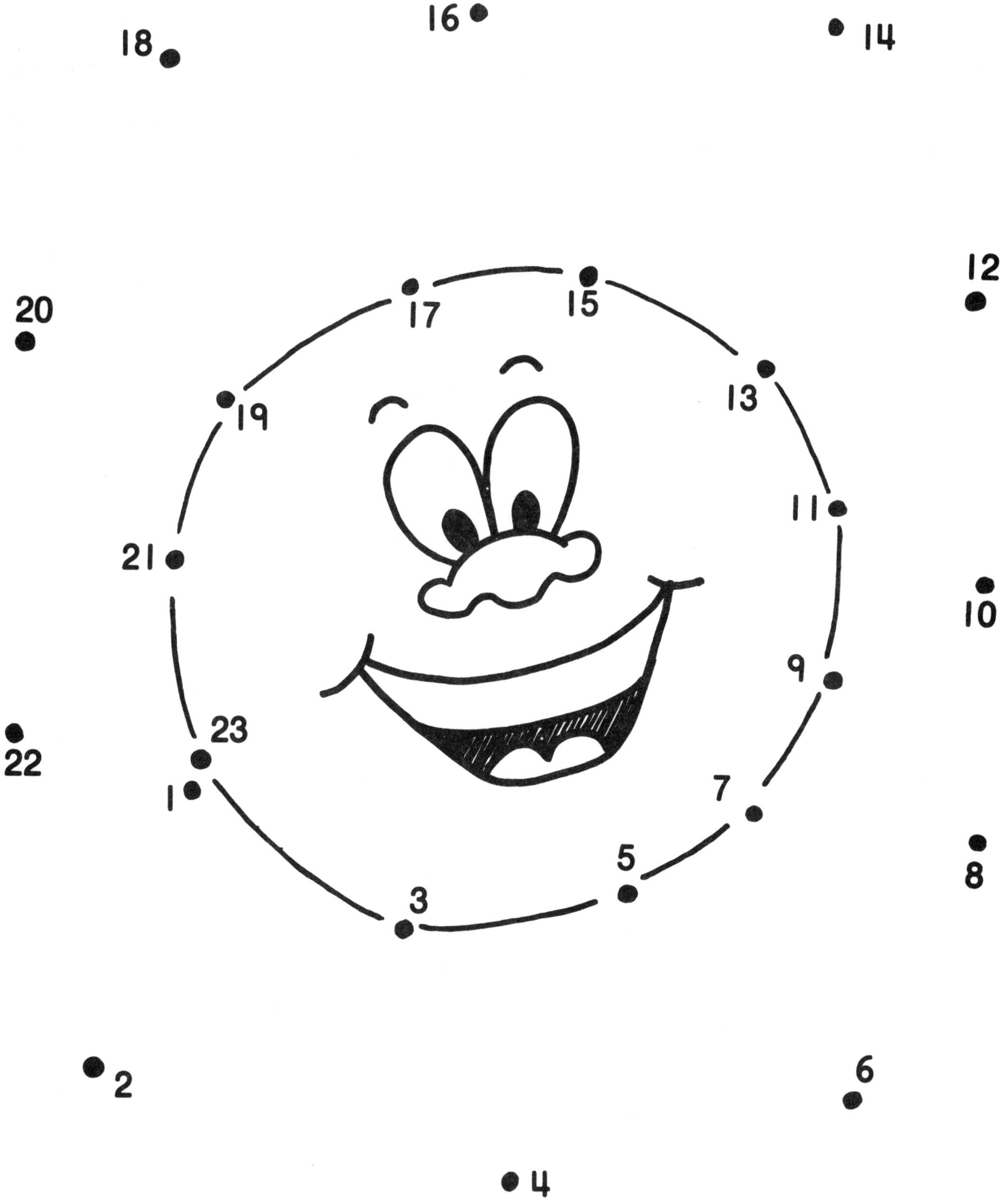

Which star is the closest one?
The answer is the ____________.

Sun

This book belongs to ______________________________

Name ______________________ Read, trace and color.

The Sun Is the Center of the Solar System

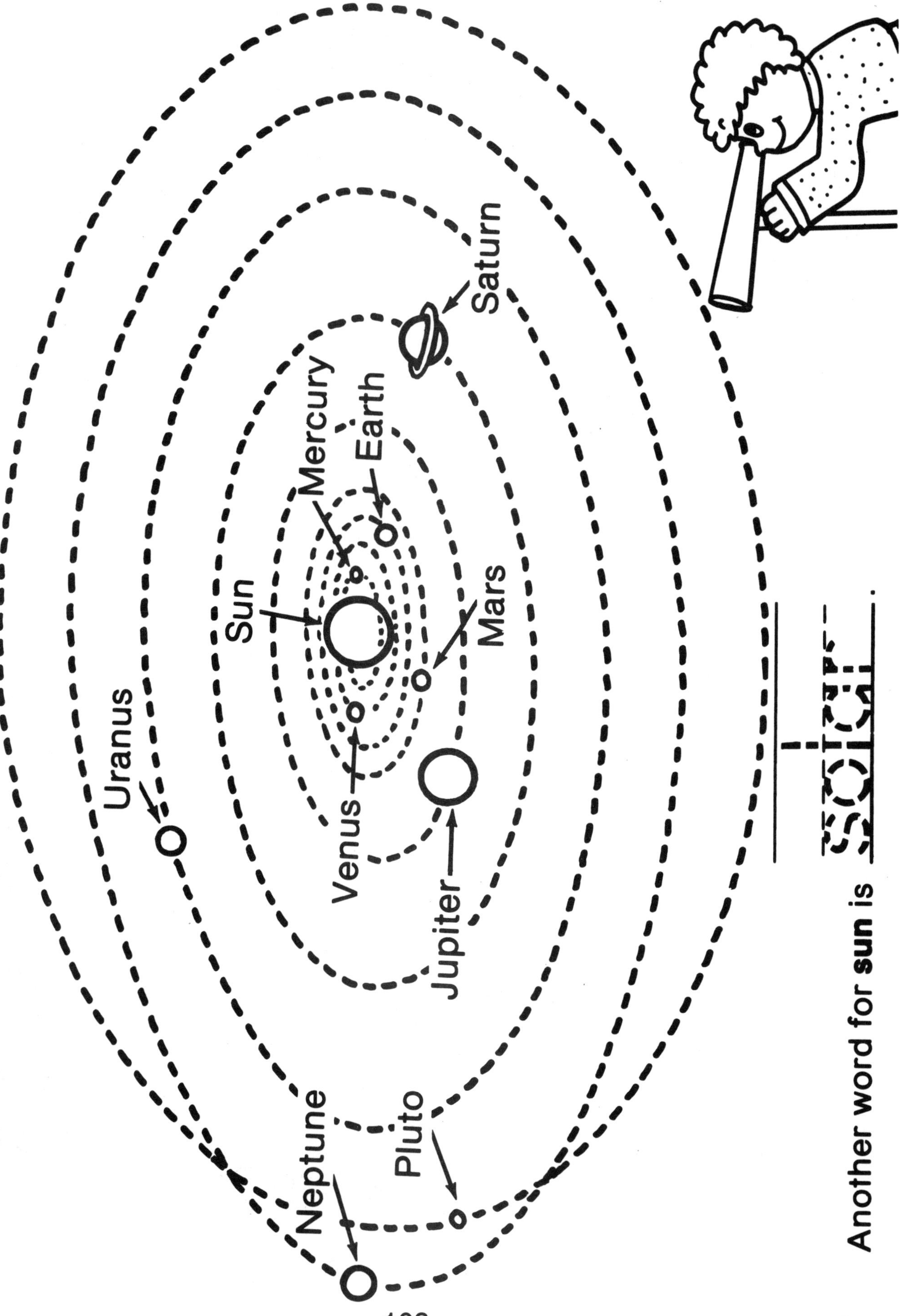

Another word for **sun** is solar.

FS-32047 Science

Name ______________________ Color using the color words.

The Sun Is the Star Nearest the Earth

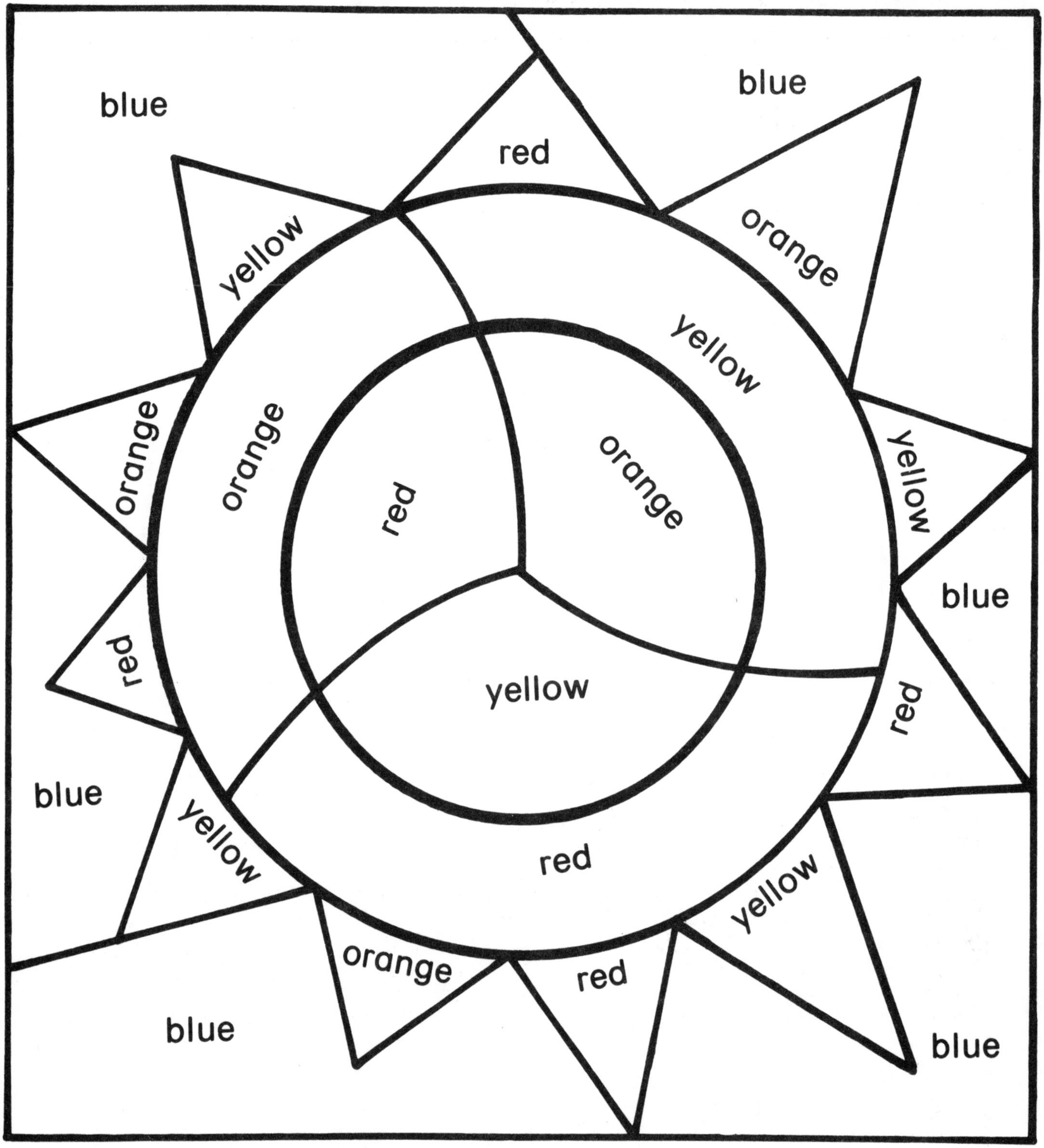

____

- - - - - - - - - - - -

The ______________________ is the only daytime star.

Name ______________________________ Cut and paste in the correct place.

We See Sunlight in the Daytime

Daytime		Nighttime	

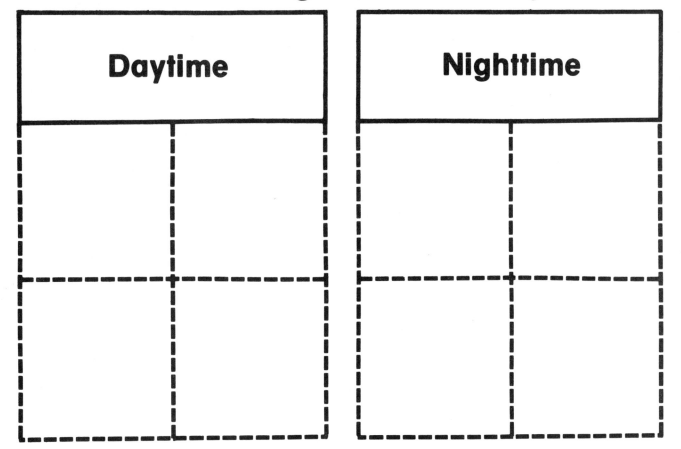

Name ______________________ Color using the code.

The Sun Makes Plants Grow

green

orange

purple

yellow

Name ______________________________ Cut and paste the boxes.

The Sun Gives Us Heat and Light

1.
2.
3.
4.
5.

Name ______________________________ Match the shadows.

Sunlight Makes Shadows

Stars

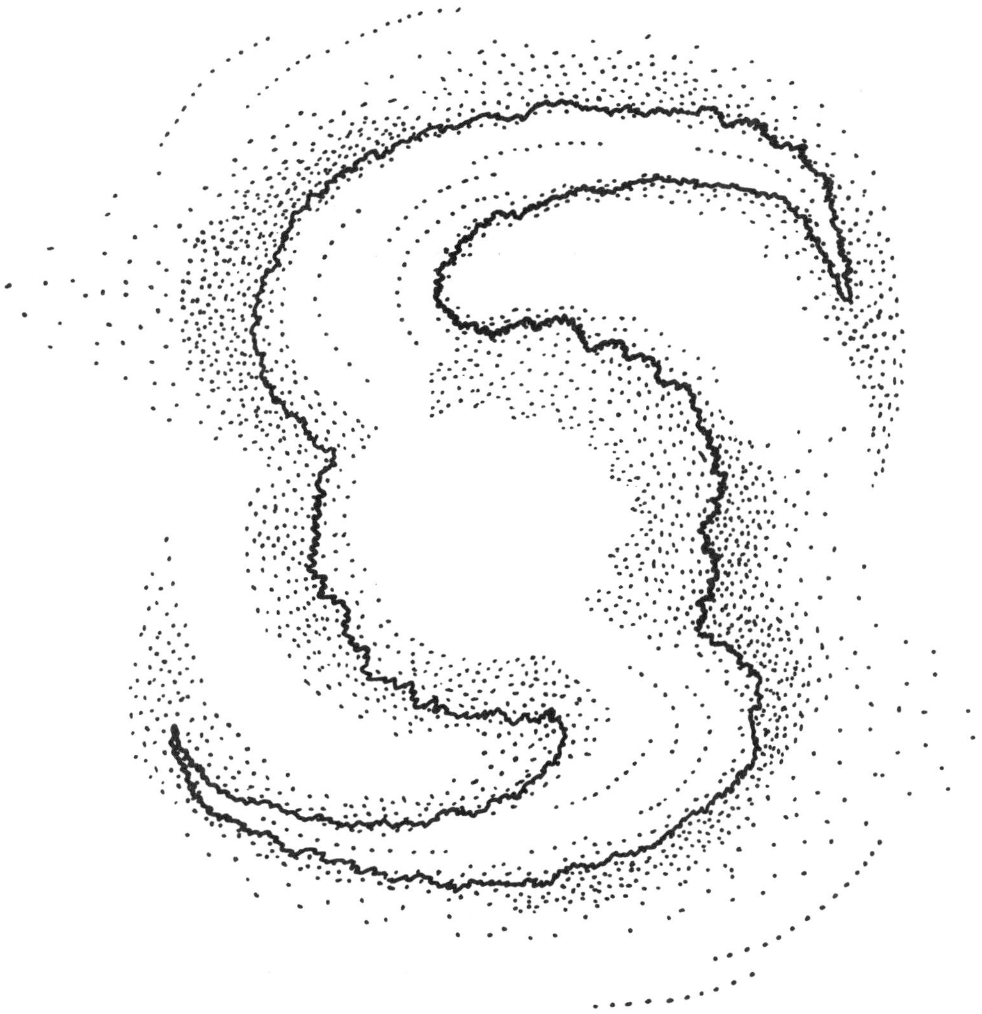

The Milky Way Galaxy

This book belongs to ______________________

Name ______________________________ Color and remember.

The sun is our day star.

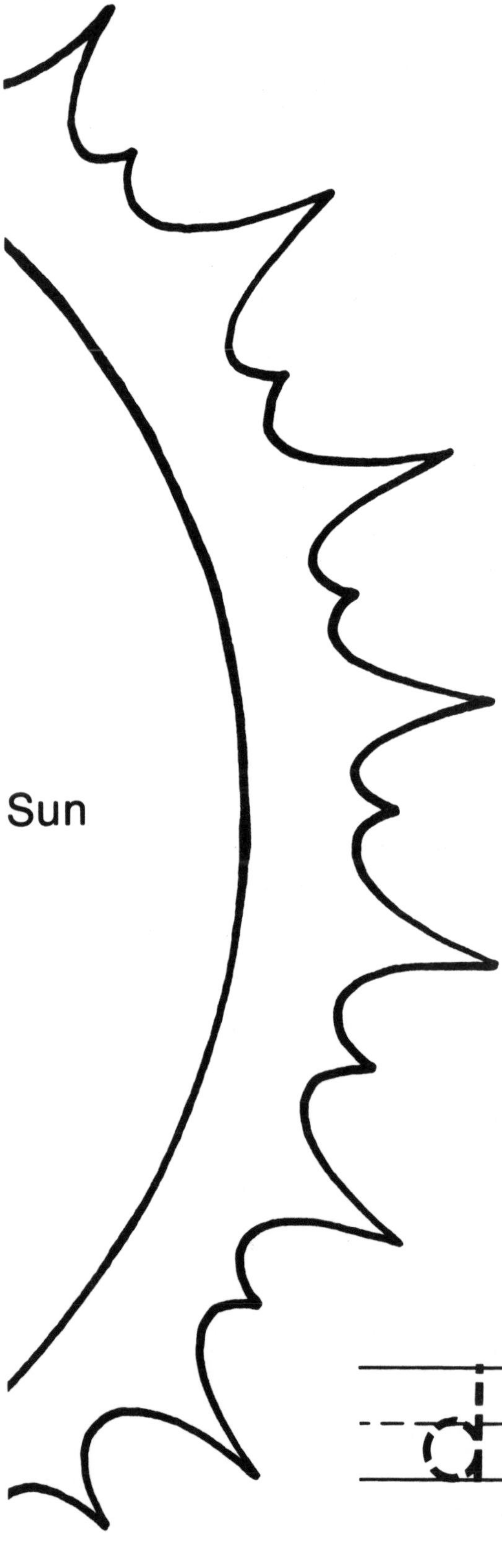

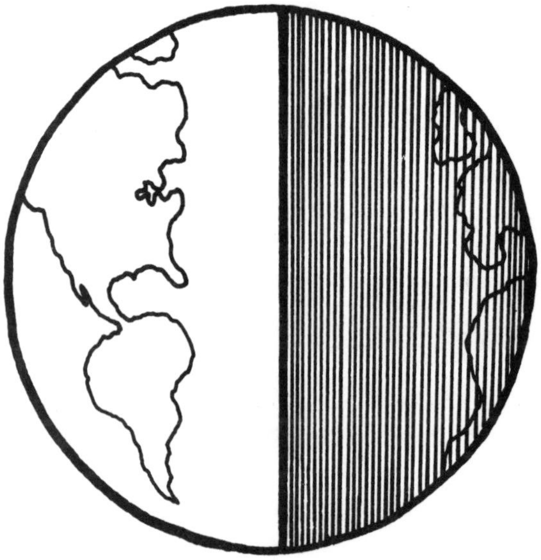

Earth

Day **Night**

day star

Name ______________________________ Connect the stars.

Constellations are groups of stars.

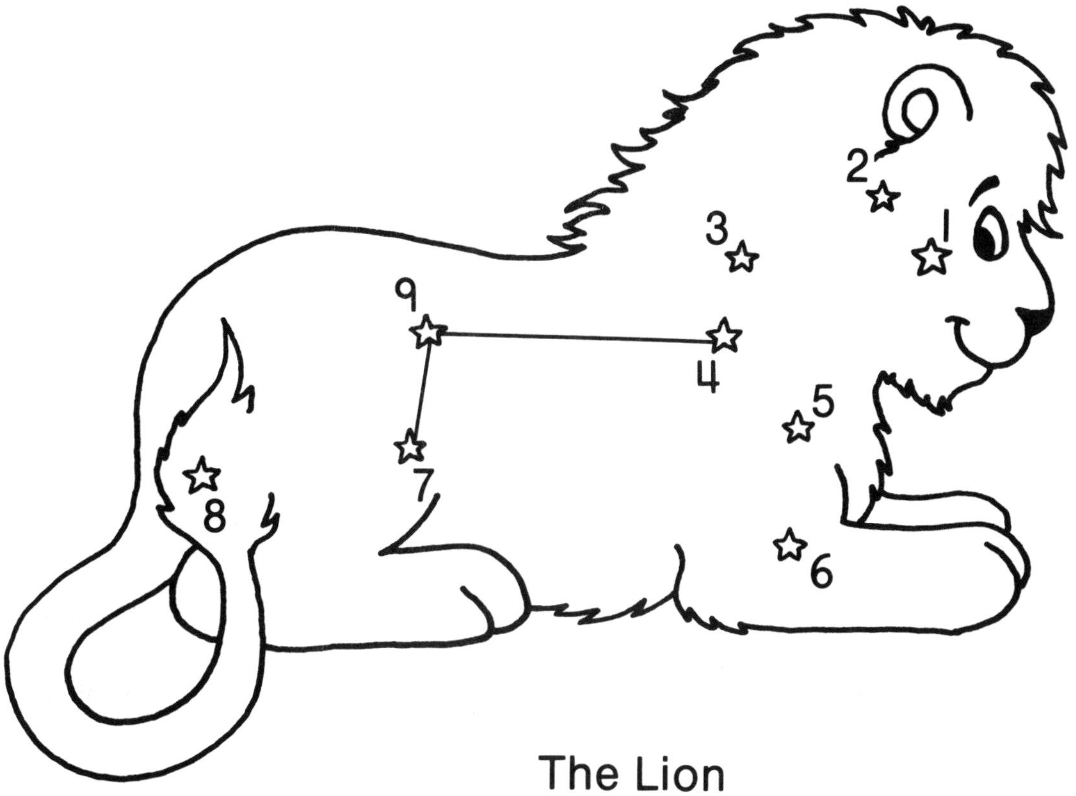

The Lion

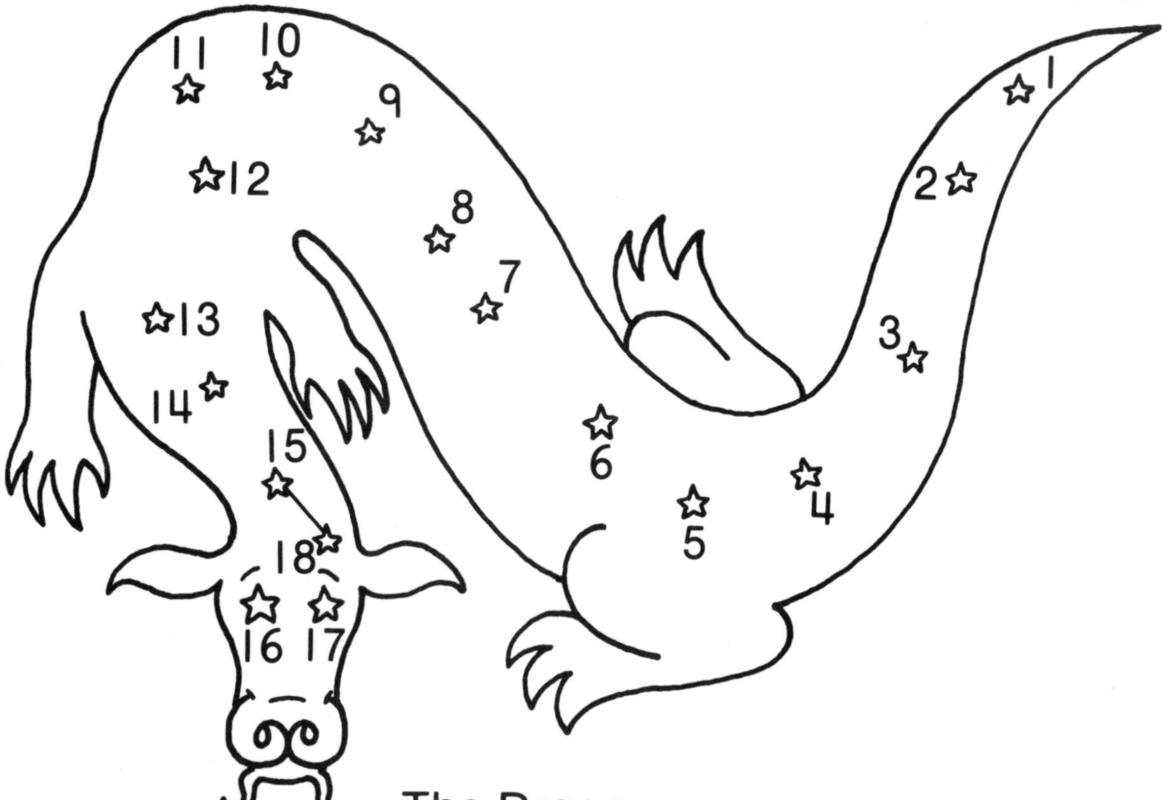

The Dragon

Name ________________________________ Cut and paste.

The North Star helps us find the way.

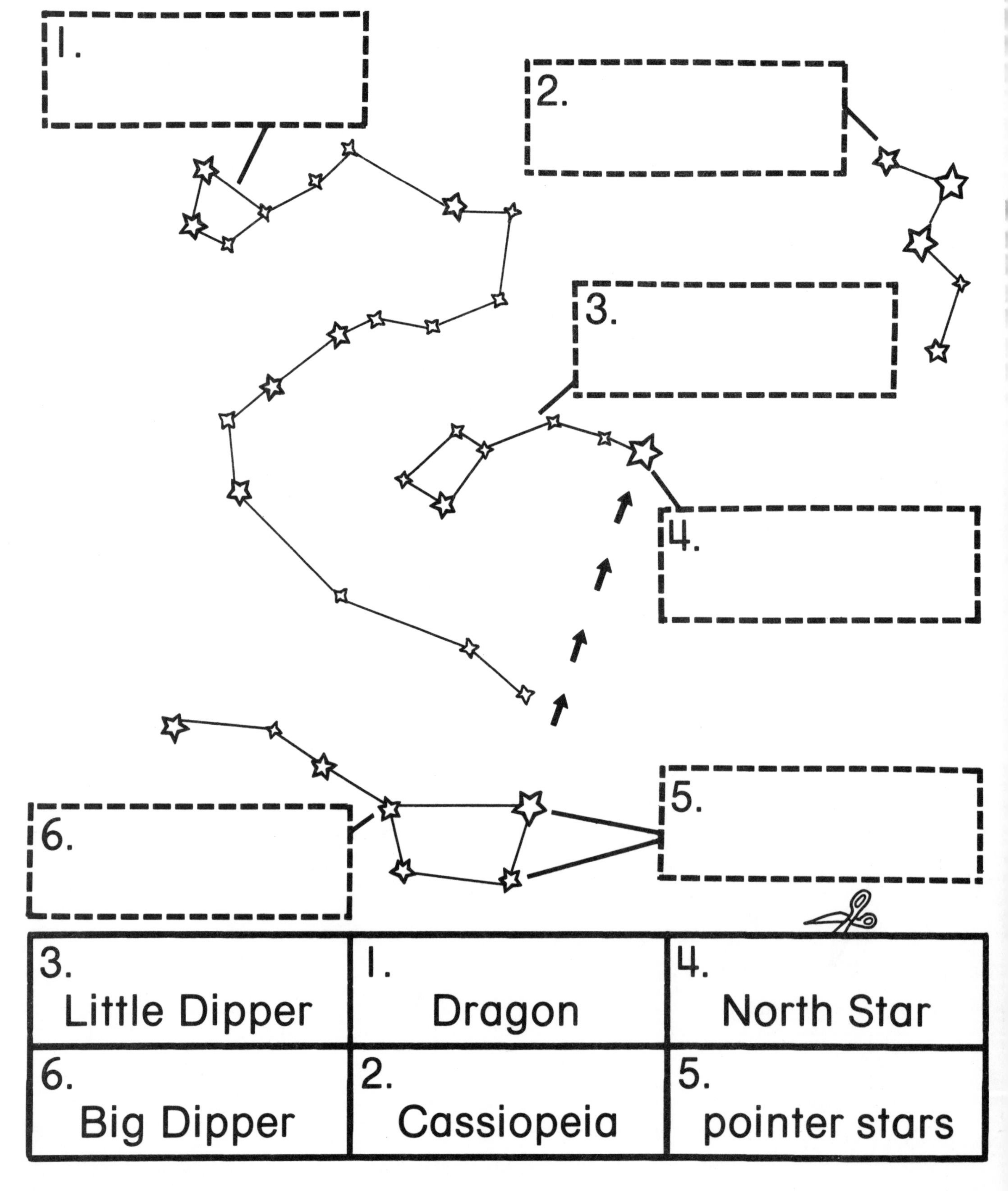

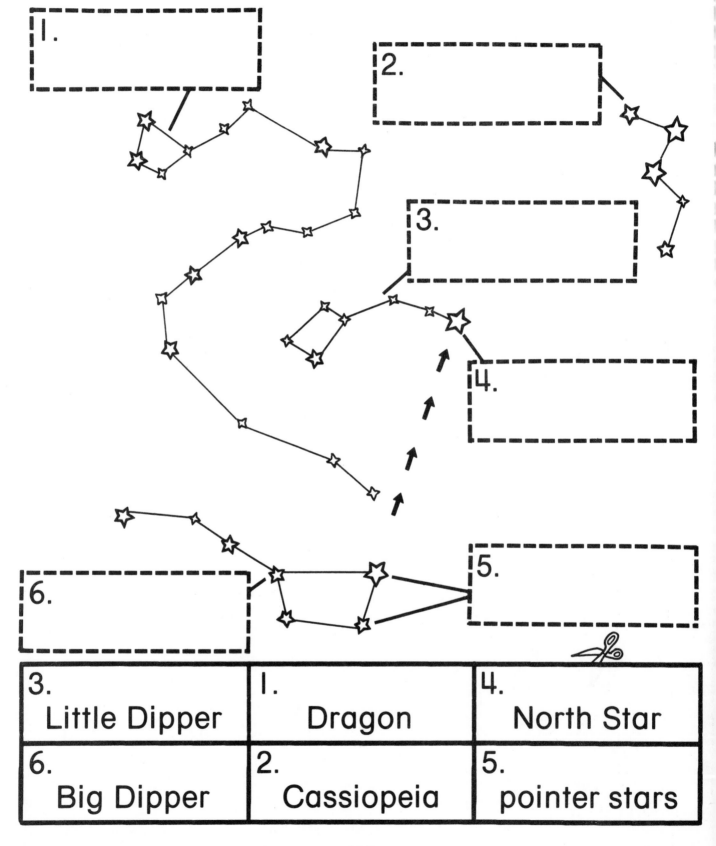

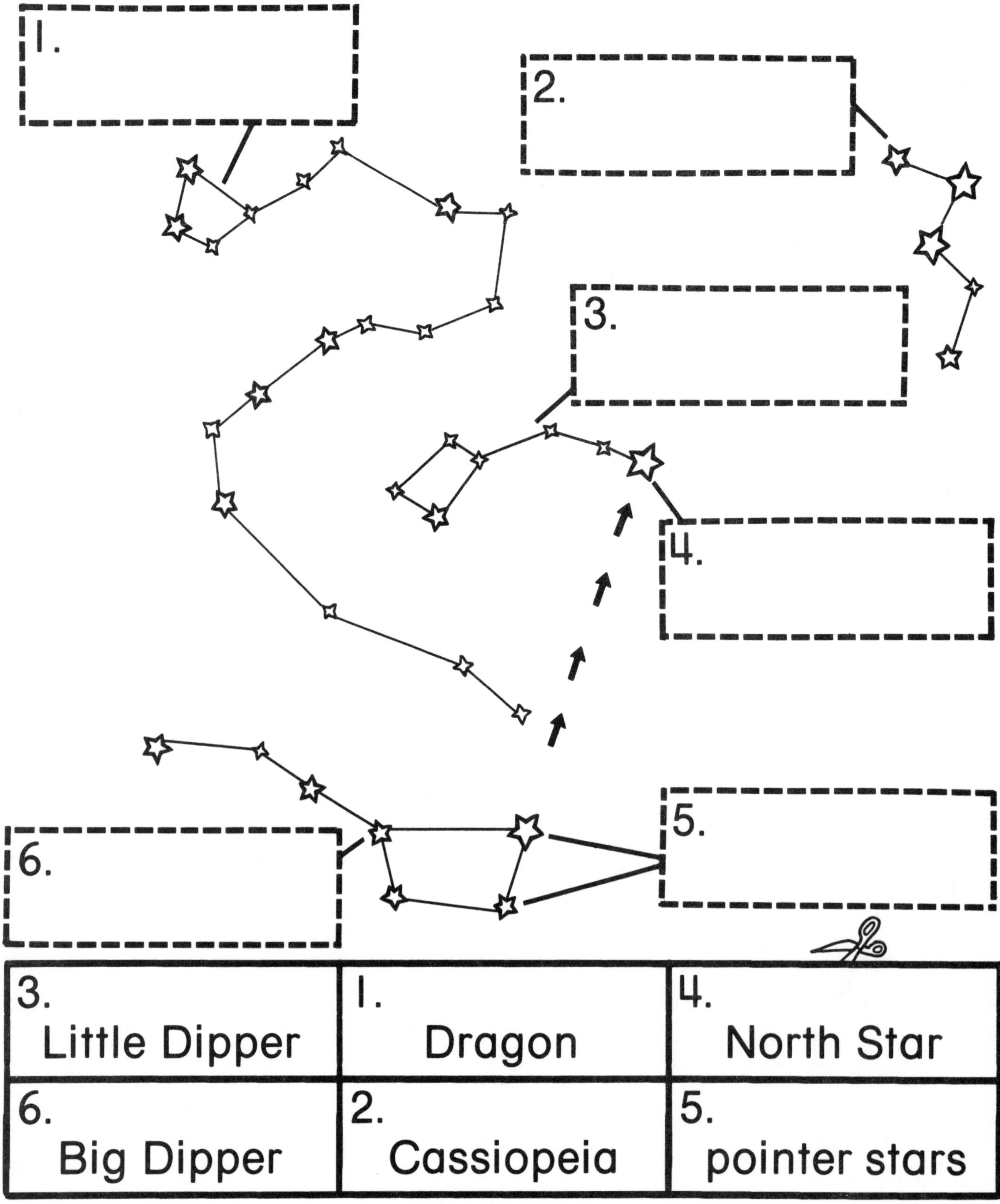

3. Little Dipper	1. Dragon	4. North Star
6. Big Dipper	2. Cassiopeia	5. pointer stars

Name ______________________________ Trace and color.

Stars Have Colors

Cool star

red

orange

yellow

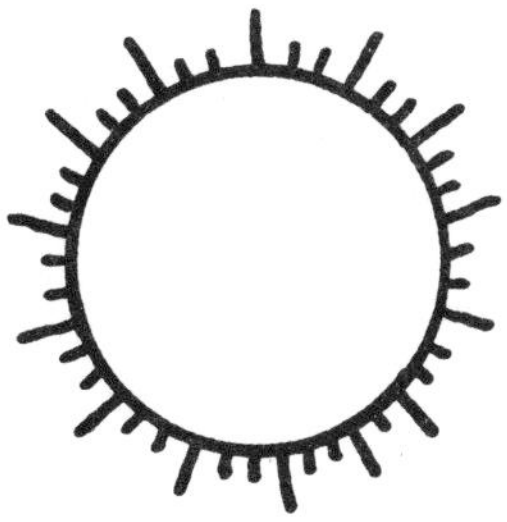

white

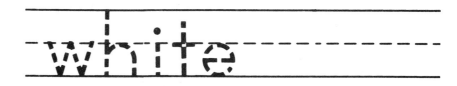

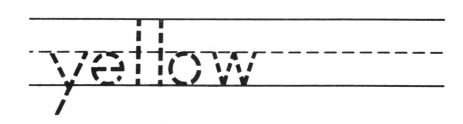

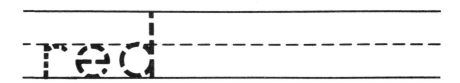

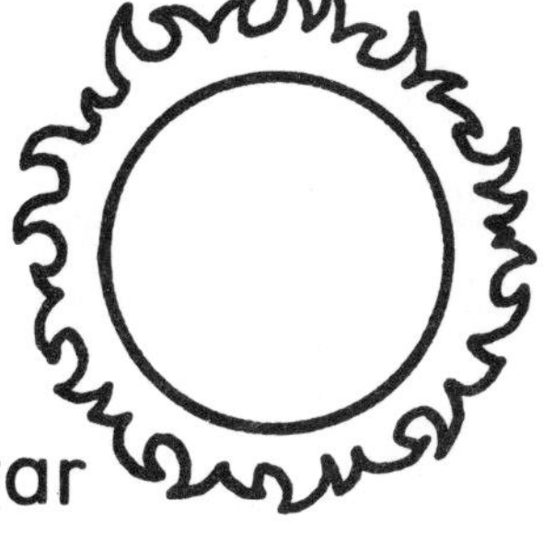

blue

Hot star

Name ______________________________ Cut and paste.

Look for Stars at Night

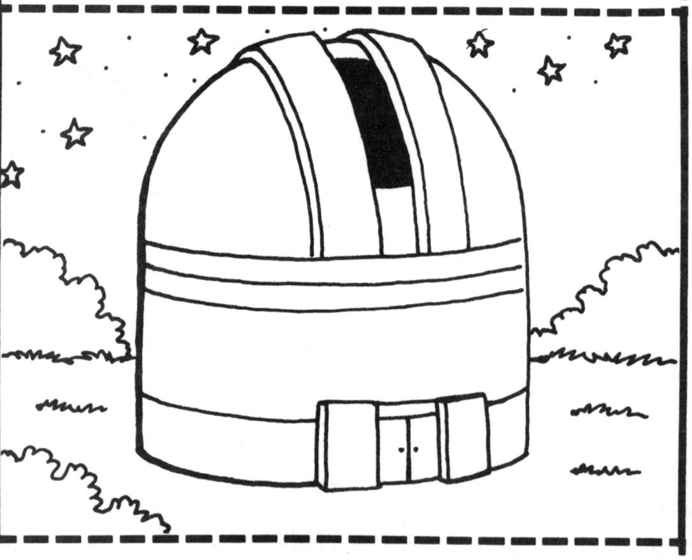

in the mountains	away from lights
from an observatory	with a telescope

Water

This book belongs to ______________________________

Name ______________________________ Color and trace.

Water is a liquid.

liquid

Name ______________________ Connect the dots from A-Z.

Water is a solid.

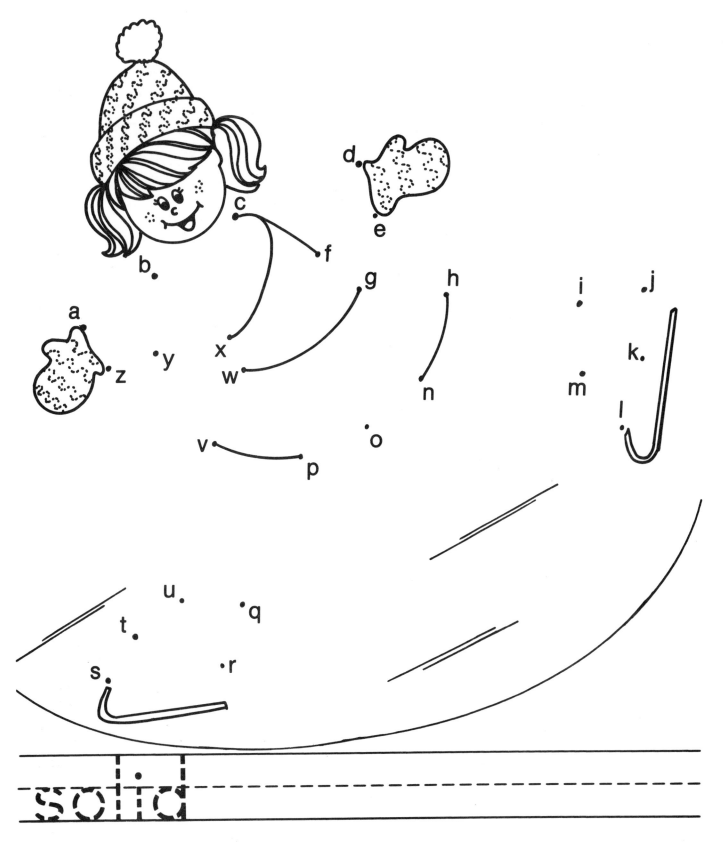

solid

Name ______________________________ Match the containers.

Water is a gas.

gas

Name ______________________ Color, cut and paste.

Water can be —

solid

gas

liquid

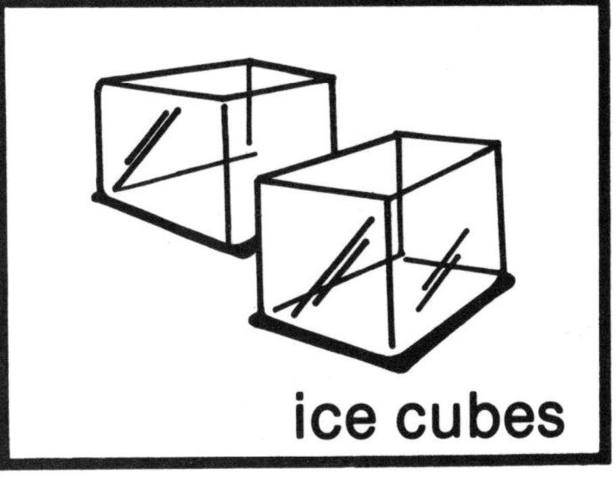

Name ______________________________ Match the pictures.

All living things need water.

Name ______________________ Color the water cycle.

Water changes.

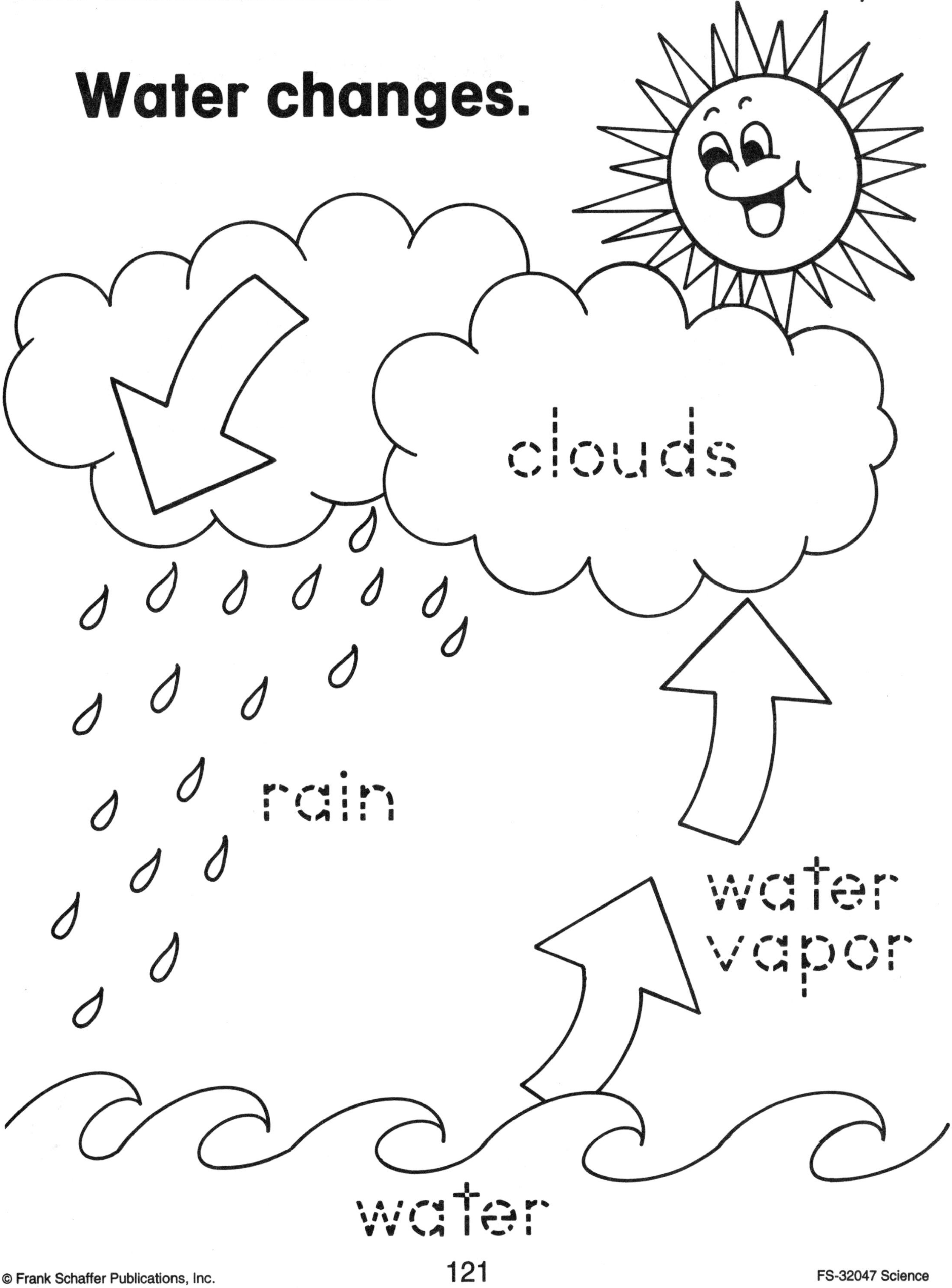

Oceans

This book belongs to ________________

Name ______________________________ Color the animals that live in the ocean.

Ocean Animals

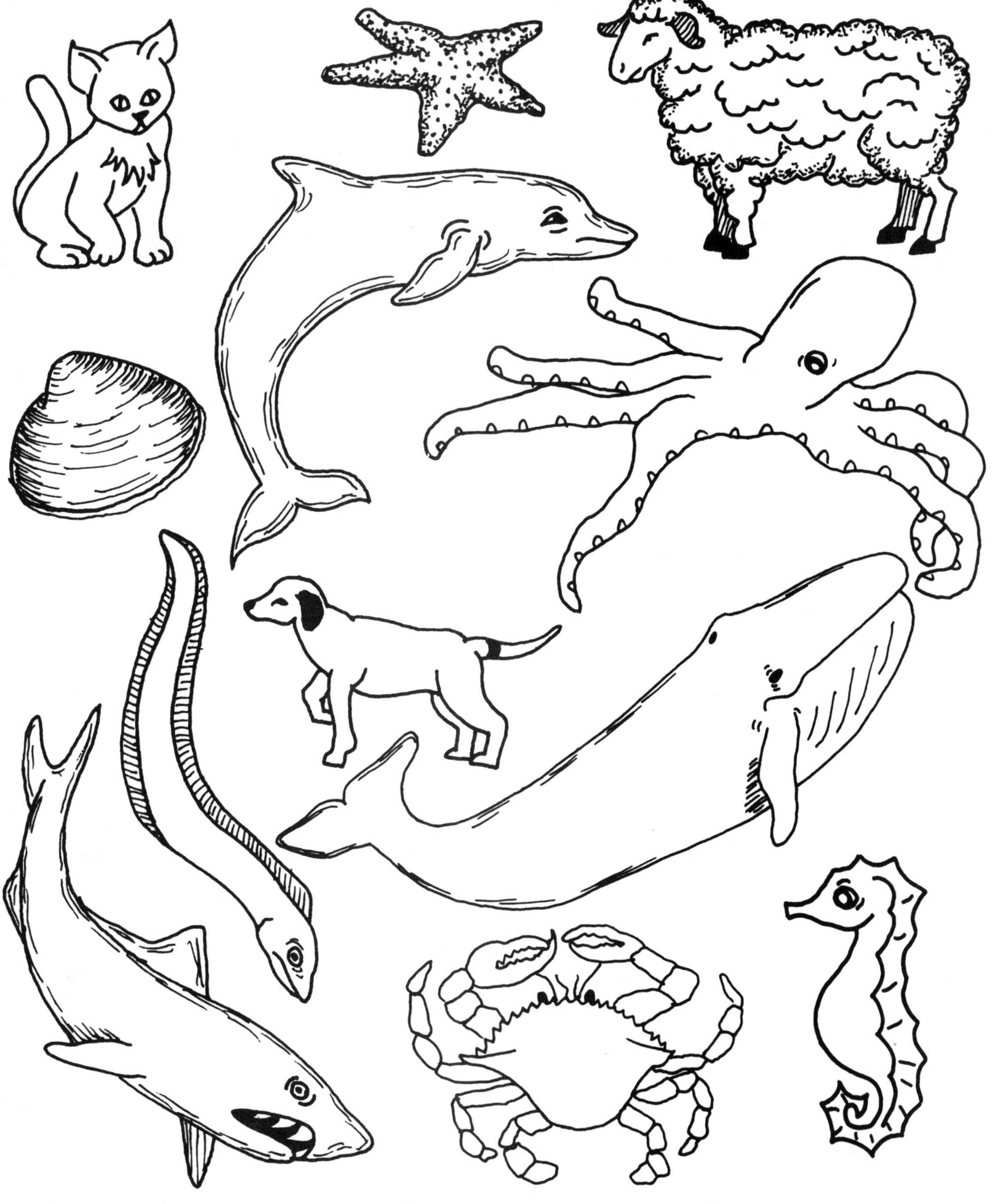

Name ____________________ Color, cut and paste to complete the fish.

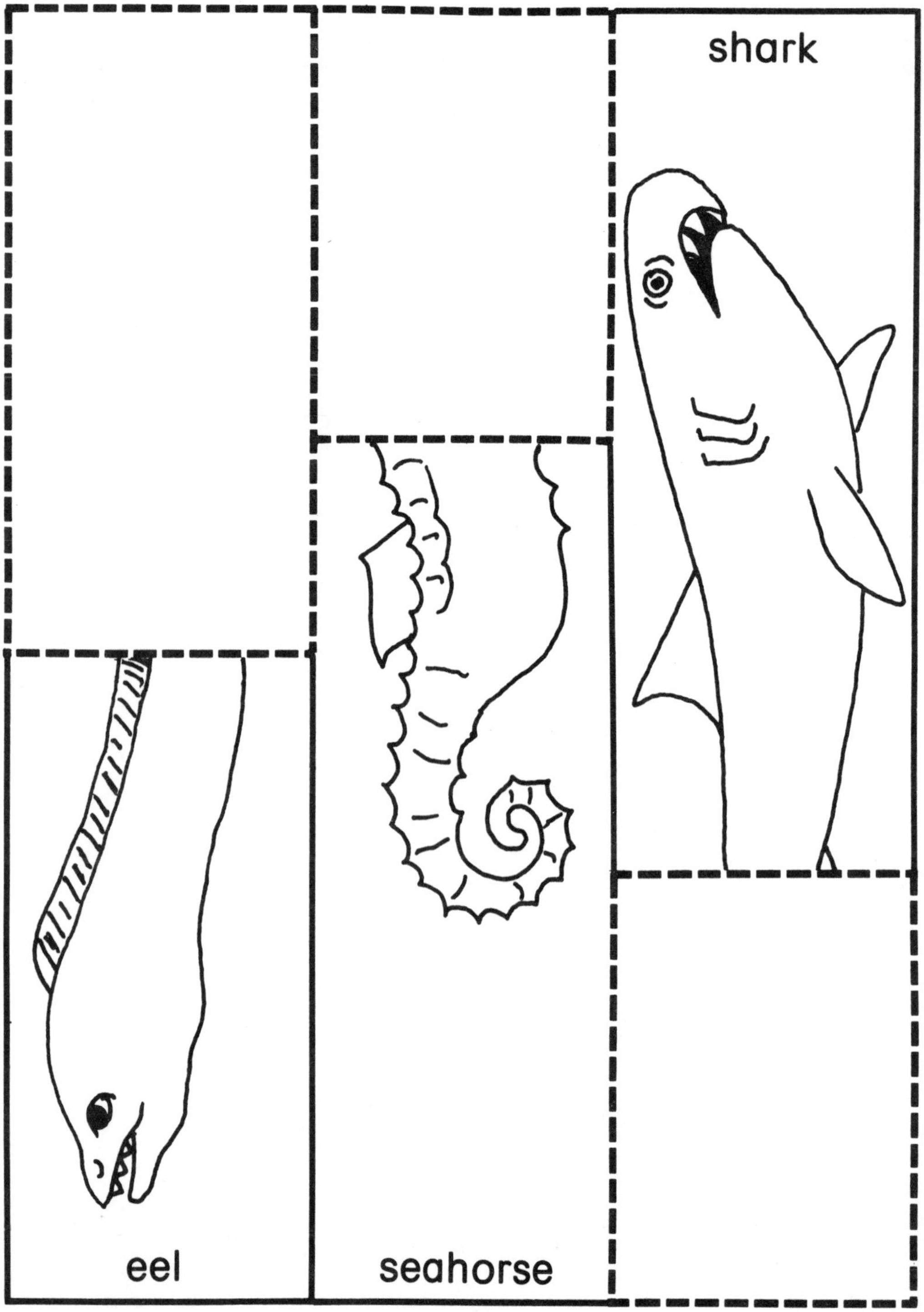

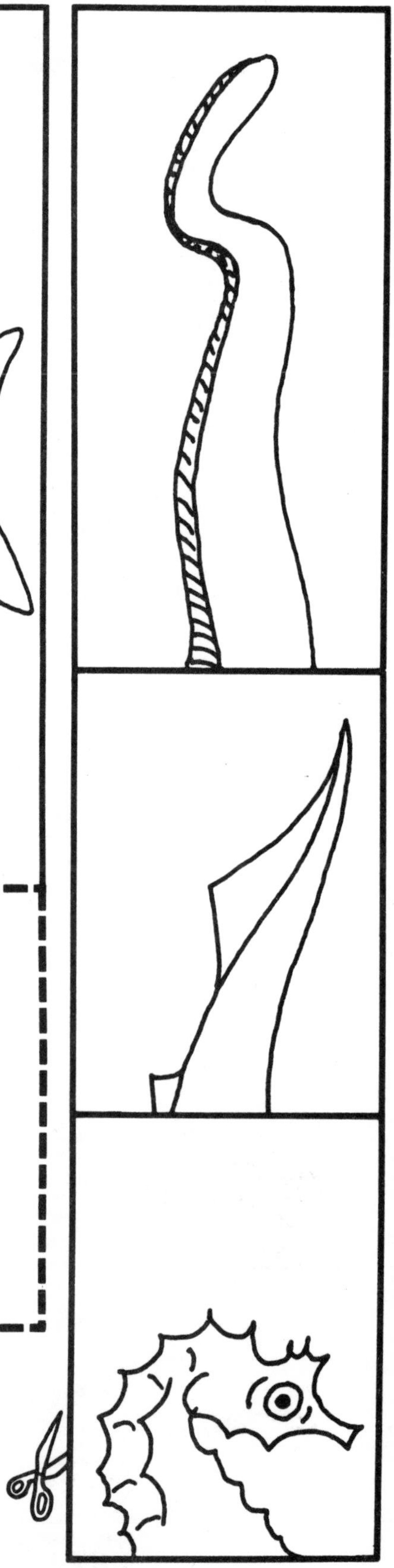

Fish breathe underwater.

Name ______________________________ Read and color.

green red purple orange

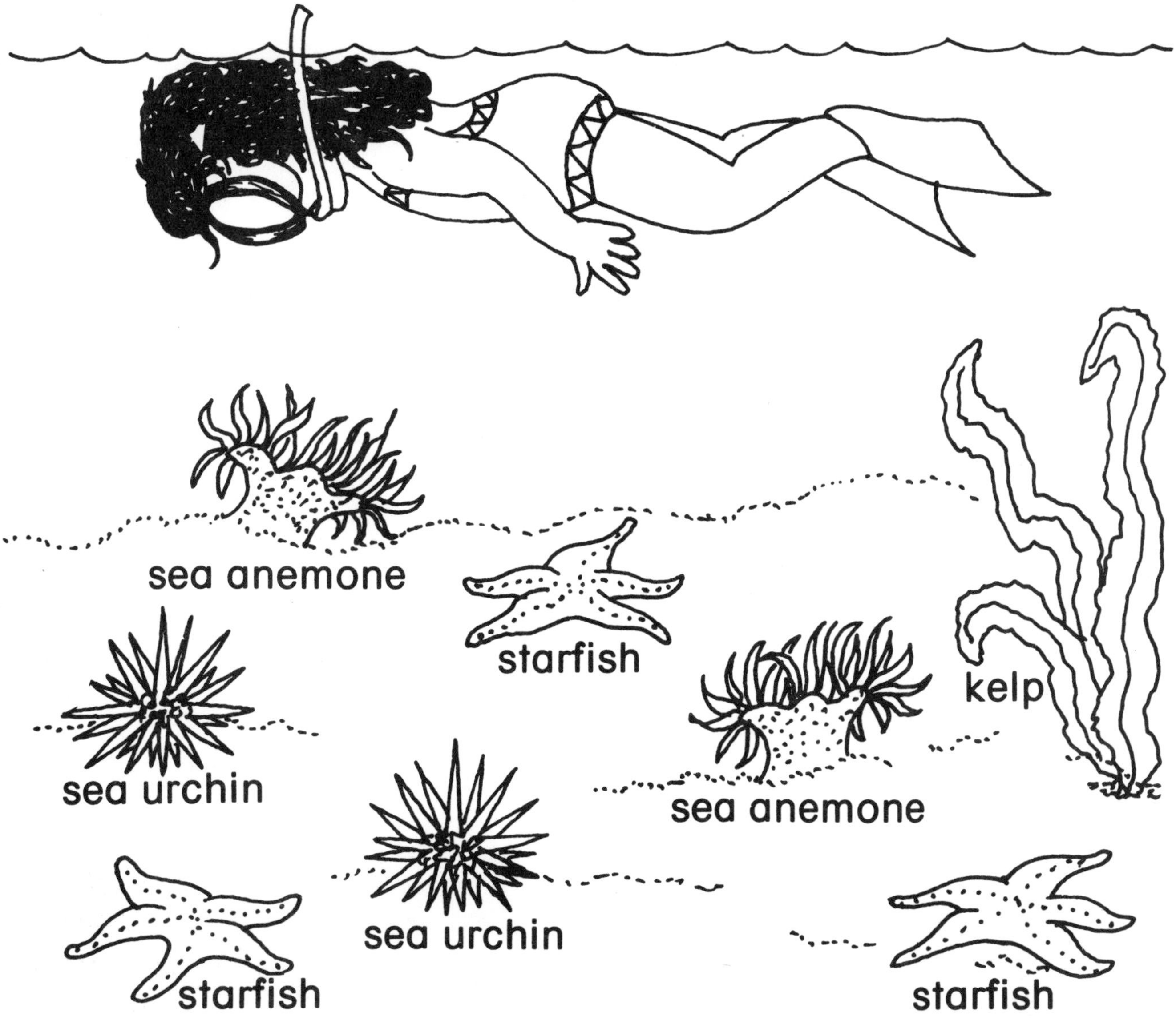

What lives underwater?

Name________________________________Cut and paste in order.

Blue whales are the largest animals on earth.

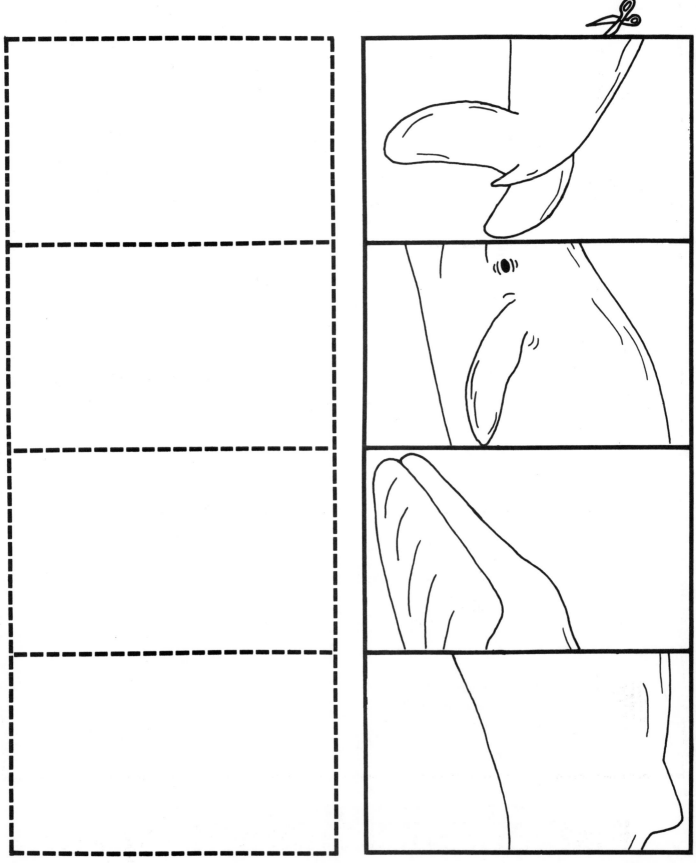

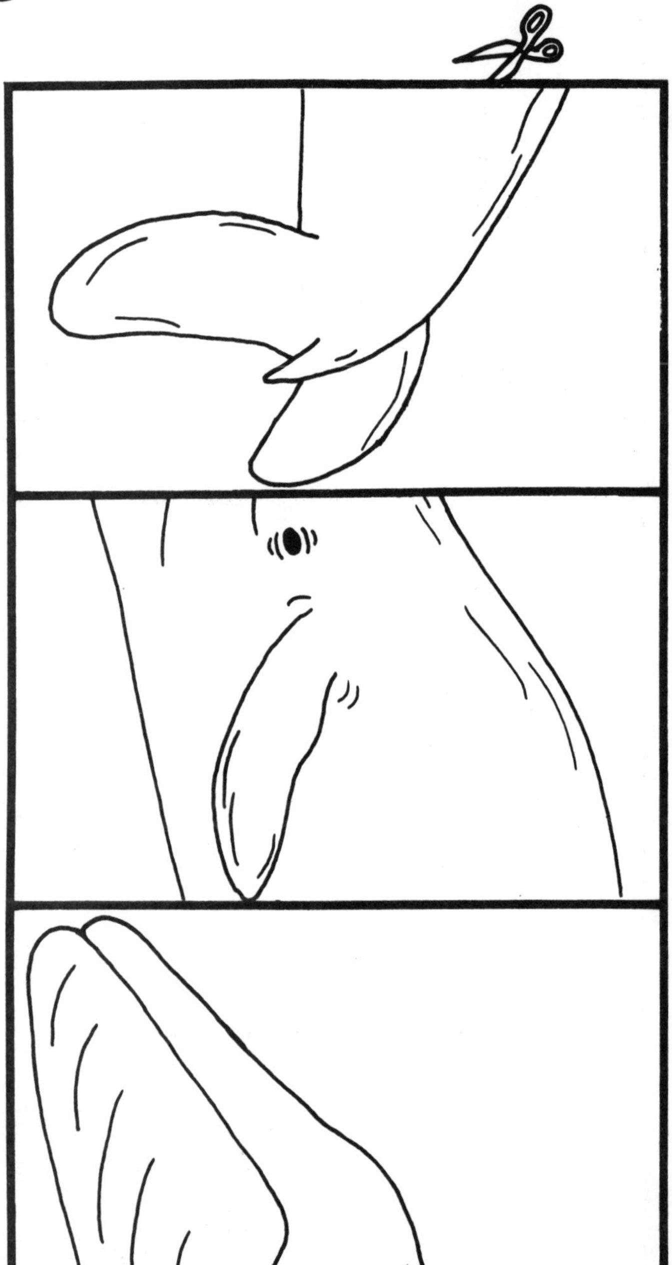

Name ______________________ Complete each row. Color.

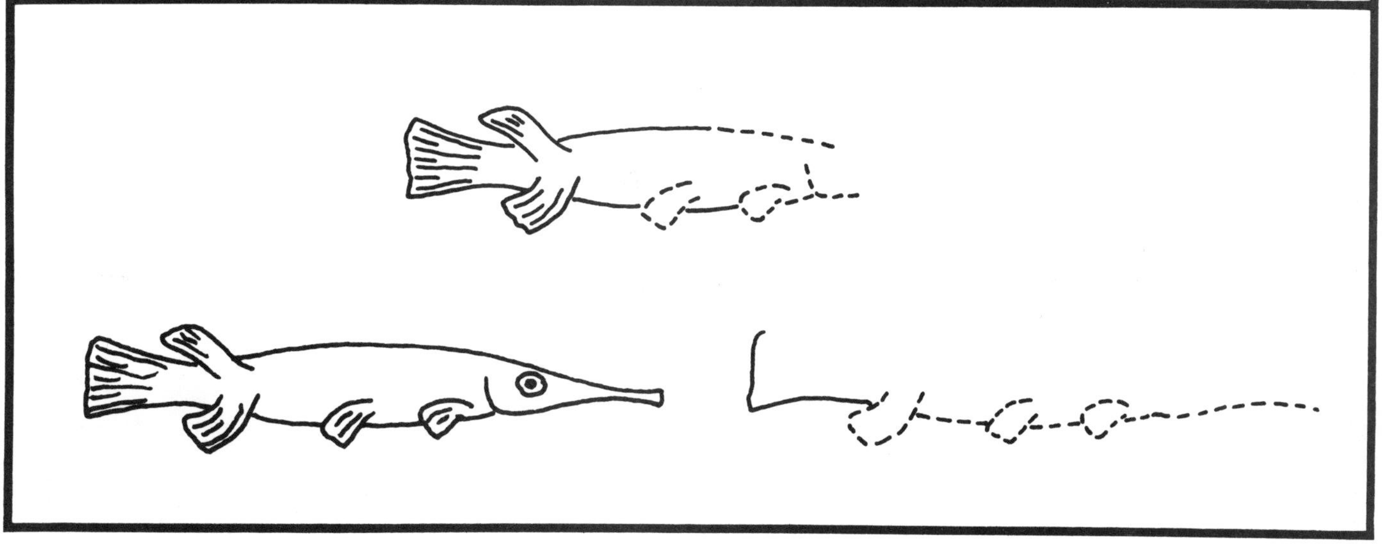

Fish come in many shapes.

Name ______________________

Color. Cut and paste in the correct place.

Sea Animals	Farm Animals